Nail your Maths exams with CGP!

There are some things in life that really shouldn't be left to chance —
making cheese on toast, your hairdo or the SQA Higher Maths exams.
For the latter, you'll need plenty of practice to be totally prepared.

That's why the experts at CGP have created this fantastic Workbook.
It's bursting with exam-style questions covering the SQA course topic-by-topic
— plus a full set of practice papers that are so like the real exams, it's scary.

We've even included step-by-step answers and mark schemes at the back,
so you'll be a Higher Maths pro in absolutely no time!

CGP — the best by miles! ☺

Our sole aim here at CGP is to produce the highest quality books —
carefully written, immaculately presented and dangerously close to being funny.

Then we work our socks off to get them out to you
— at the cheapest possible prices.

Published by CGP

Editors:
Sammy El-Bahrawy, Sarah George, Shaun Harrogate, Samuel Mann, Michael Weynberg

ISBN: 978 1 78294 961 9

With thanks to Liam Dyer and Garry Simpson for the proofreading.

Clipart from Corel®
Printed by Elanders Ltd, Newcastle upon Tyne

Based on the classic CGP style created by Richard Parsons.

Contents

✓ Use the tick boxes to check off the topics you've completed.

Exam Tips

Exam Stuff

1) You will have <u>two</u> exams — one <u>non-calculator</u> exam and one <u>calculator</u> exam.

2) The non-calculator exam is <u>1 hr 30 mins</u> long and worth <u>70 marks</u>.

3) The calculator exam is <u>1 hr 45 mins</u> long and worth <u>80 marks</u>.

4) Timings in the exam are really important, so here's a quick guide...

- You should spend roughly 1 minute per mark working on each question (4 marks ≈ 5 minutes).
- Then, if you've got any time left at the end of the exams, check back through your answers and make sure you haven't made any silly mistakes. Don't just stare at that hottie in front.
- If you're totally, hopelessly stuck on a question, just leave it and move on to the next one. You can always go back to it at the end if you've got enough time.

There are a Few Golden Rules

1) **Always, always, always make sure you <u>read the question properly</u>.**
   For example, if the question asks you to give your answer in metres, <u>don't</u> give it in centimetres.

2) **Show <u>each step</u> in your <u>working</u>.**
   You're less likely to make a mistake if you write things out in stages. And even if your final answer's wrong, you'll probably pick up <u>some marks</u> if the examiner can see that your <u>method</u> is right.

3) **Check that your answer is <u>sensible</u>.**
   It's no good giving an answer of 45° if the question asks for solutions between 180° and 360°.

4) **Make sure you give your answer to an appropriate <u>degree of accuracy</u>.**
   The question might tell you how to round your answer, but if it doesn't then just choose something sensible. Either way, don't round anything until the end of the calculation.

5) **Look at the number of <u>marks</u> a question is worth.**
   If a question's worth 2 or more marks, you're not going to get them all for just writing down the final answer — you're going to have to <u>show your working</u>.

6) **Write your answers as <u>clearly</u> as you can.**
   If the examiner can't read your answer you won't get any marks, even if it's right.

Obeying these Golden Rules will help you get as many marks as you can in the exam — but they're no use if you haven't learnt the stuff in the first place. So make sure you revise well and do <u>as many</u> practice questions as you can.

Using Your Calculator

1) Your calculator can make questions a lot easier for you, but only if you <u>know how to use it</u>. Make sure you know what the different buttons do and how to use them.

2) Remember to check whether your calculator needs to be in <u>degrees mode</u> or <u>radians mode</u> before tackling the question. This is important for <u>trigonometry</u> questions.

3) If you're working out a <u>big calculation</u> on your calculator, it's best to do it in <u>stages</u> and use the <u>memory</u> to store the answers to the different parts. If you try and do it all in one go, it's too easy to mess it up.

4) If you're going to be a renegade and do a question all in one go on your calculator, use <u>brackets</u> so the calculator knows which bits to do first.

REMEMBER: <u>Golden Rule number 2</u> still applies, even if you're using a calculator — you should still write down <u>all</u> the steps you are doing so the examiner can see the method you're using.

Reasoning Skills

Reasoning skills are all about how you tackle a mathematical problem and explain the answer.
You'll have to use reasoning skills on both exam papers, so it's a good idea to know what you need to do.

There are Two Main Types of Reasoning Questions...

1) For some questions, it's <u>obvious</u> what you have to do — e.g. 'differentiate the function f(x) = x^2 + 5x' is clearly a question on <u>differentiation</u>, so solve it using the usual methods.

2) However, in other questions, it's <u>not obvious</u> what you have to do — you're given some <u>information</u>, and have to work out what <u>methods</u> to use to answer the question. These questions are designed to assess how you approach an <u>unfamiliar problem</u> — they test your <u>reasoning skills</u>.

3) Reasoning questions come in <u>two different forms</u> — in both cases, you have to work out what maths to do <u>for yourself</u>.

Questions With a Real-Life Context

1) If questions have a <u>real-life context</u>, the information you need might be <u>hidden</u>. Read through the question and work out which bits are <u>relevant</u> to the maths, and which bits are just <u>setting the scene</u>.

2) Once you've worked this out, decide which <u>method(s)</u> you need to use to answer the question. After doing all your <u>calculations</u>, make sure you <u>link</u> your answer back to the <u>original context</u>.

3) It can be pretty obvious what maths you need to use, even with a real-life context — <u>differentiation</u>, <u>exponential</u> and <u>trigonometry</u> problems are often given in context. Sometimes it's <u>not</u> that obvious though — which leads us nicely onto...

Sometimes questions will be a mixture of <u>both</u> — i.e. problem-solving questions set in a real-life context.

Problem-Solving Questions

1) In <u>problem-solving questions</u>, you'll be given a load of <u>mathematical information</u> (sometimes including <u>diagrams</u>) and asked to '<u>calculate</u>', '<u>find</u>' or '<u>determine</u>' something. It's then up to you to come up with a <u>strategy</u> to answer the question.

2) You won't be given any <u>guidance</u> for what method to use — you have to work it out for yourself.

3) There are often a couple of <u>different ways</u> you could answer the question — and you'll get the marks whichever way you do it, as long as you get the <u>answer</u> right and <u>show your working</u> clearly.

4) Once you think you've done all of the maths required, <u>check</u> you've actually answered the <u>question</u> — you might need to <u>explain</u> why your solution is the correct one, or link your answer back to a <u>context</u>.

Here are Some Useful Tips for Reasoning Questions

Unfortunately, there's <u>no</u> one set method for answering reasoning questions — they can be on anything the examiners fancy, so will involve <u>different bits</u> of maths. These <u>tips</u> should help you get started though. Don't forget the Golden Rules from the previous page. They all still apply to reasoning questions.

- <u>Read the question</u> two or three times and work out what you're <u>trying to find</u>.
- Write down what you <u>know</u> — pick out any <u>numbers</u> given in the question, and add <u>labels</u> to diagrams if you can. If you're not given a diagram, it's often a good idea to <u>sketch one yourself</u>.
- See if anything <u>jumps out</u> at you — for example, a diagram of a circle might mean you need to use the equation of the circle $(x - a)^2 + (y - b)^2 = r^2$, or mention of maximums / minimums might mean you need to differentiate to find these values.
- <u>Don't rush</u> into a problem-solving question — <u>take your time</u> and <u>think it through</u> first. Make sure you have an <u>idea</u> of what you're going to do before diving in.

Quadratic Equations

It's really important that you're confident with solving quadratic equations, as they come up all the time.
That includes the different ways of solving them — completing the square, factorising and the quadratic formula.

1   How many roots does the equation $x^2 - 4x + 2 = 0$ have?

....

(2 marks)

2   Given that $5x^2 + nx + 14 \equiv m(x + 2)^2 + p$, find the values of the integers m, n and p.

$m =$ .... $n =$ .... $p =$ ....

(3 marks)

3   Express $x^2 - 18x + 16$ in the form $(x + a)^2 + b$.

....

(2 marks)

4   The equation $ax^2 + 2x + 2 = 0$ has one real root.
What is the value of a?

$a =$ ....

(2 marks)

Quadratic Equations

5 The equation $x^2 - 4x + (k-1) = 0$, where k is a constant, has no real roots.
Find the range of possible values of k.

....

(3 marks)

6 $f(x) = \dfrac{1}{x^2 - 7x + 17}$

a) Express $x^2 - 7x + 17$ in the form $(x - m)^2 + n$, where m and n are constants.

....

(2 marks)

b) Hence find the maximum value of $f(x)$.

....

(2 marks)

7 Find the possible values of k if the equation $g(x) = 0$ is to have one real root,
where $g(x)$ is given by $g(x) = 3x^2 + kx + 2k$.

....

(3 marks)

Score

19

Quadratic Inequalities

Now that you're up to speed with quadratic equations, it's time to move on to solving quadratic inequalities. Keep the inequality sign pointing the same way and only flip it when you multiply or divide by a negative number.

1 Solve the inequality $3x^2 - 5x - 2 \leq 0$.

....
(2 marks)

2 A rectangular office is to be built, measuring $(x - 9)$ metres wide and $(x - 6)$ metres long. Given that at least 28 m² of floor space is required, find the set of possible values of x.

....
(3 marks)

3 Look at the two cuboids below. The volume of cuboid A is greater than the volume of cuboid B.

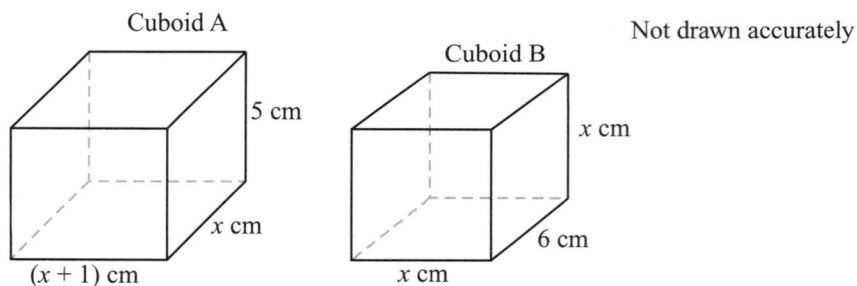

Cuboid A

Cuboid B

Not drawn accurately

5 cm

x cm

x cm

6 cm

x cm

$(x + 1)$ cm

x cm

a) Show that $x^2 - 5x < 0$.

....
(2 marks)

b) If x is an integer greater than 1, list the possible values of x.

....
(3 marks)

EXAM TIP
Sketching a graph may seem like a pain, but it can be very useful — you can find the range of values that you're looking for by seeing where the graph is above or below the x-axis. Besides, in the exam, you'll need to justify how you got your answer. Sketching the graph is a nice easy way to do this, and it'll also help you to see if you've accidentally gone wrong somewhere.

Score

10

Cubics and Quartics

A cubic can factorise into a maximum of three brackets and a quartic can factorise into a maximum of four brackets. If the examiners are feeling nice, they'll give you one or two factors, which makes things a bit easier.

1 For the function $f(x) = x^3 - 6x^2 - x + 30$:

   a) Use the factor theorem to show that $(x - 3)$ is a factor of $f(x)$.

   (2 marks)

   b) Factorise $f(x)$ completely.

   ....

   (2 marks)

2 Sketch each of the curves below on the axes provided.
   Show clearly any points of intersection with the x- and y-axes.

   a) $y = (x - 1)(x + 2)(3 - x)$

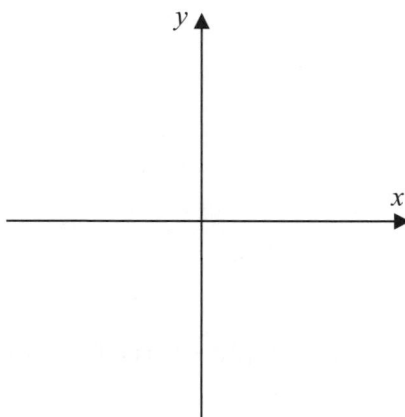

   (3 marks)

   b) $y = (x - 2)^2(x + 3)^2$

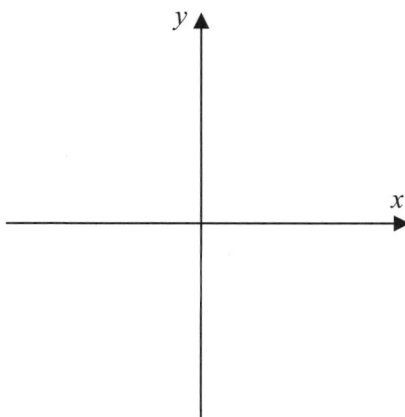

   (3 marks)

Section One — Algebraic Skills

Cubics and Quartics

3 The diagram shows the curve with equation $y = f(x)$, where $f(x) = rx(x - s)(x - t)^2$. The curve passes through $(-2, 0)$, $(0, 0)$, $(1, -4.5)$ and $(2, 0)$.

Find the values of r, s and t.

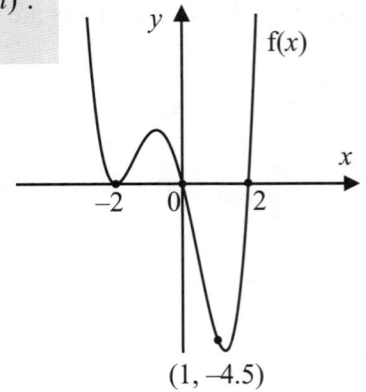

(1, –4.5)

$r =$ ...., $s =$ ...., $t =$ ....

(3 marks)

4 The function $f(x)$ is given by $f(x) = x^3 + ax^2 + bx + 7$. The remainder of $f(x)$ when divided by $(x - 3)$ is 7 and the remainder of $f(x)$ when divided by $(x + 2)$ is –3.

Find the values of a and b.

$a =$ ...., $b =$ ....

(5 marks)

5 For the function $f(x) = 2x^3 - x^2 - 5x - 2$:

a) Use the factor theorem to show that $(2x + 1)$ is a factor of $f(x)$.

(2 marks)

b) Factorise $f(x)$ completely.

....

(2 marks)

Cubics and Quartics

6 A function is defined by $f(x) = x^3 - 2x^2 - ax + 12$.
Given that $(x - 1)$ is a factor of $f(x)$, solve the equation $x^3 - 2x^2 - ax + 12 = 0$.

Use the factor theorem to find the value of a.

$x = $ ....

(5 marks)

7 The graph of the quartic function $f(x) = x^4 + ax^3 + bx^2 + cx + d$ passes through the points $(-5, 0)$, $(-3, 0)$, $(-2, 0)$, $(z, 0)$ and $(0, -120)$. Find the value of z.

$z = $ ....

(3 marks)

8 For the function $f(x) = x^4 - 2x^3 - 9x^2 + sx$:

a) Given that -24 is the remainder when $f(x)$ is divided by $(x + 1)$, show that $s = 18$.

(2 marks)

b) Given that $(x + 3)$ is a factor of $f(x)$, factorise $f(x)$ fully.

....

(3 marks)

Score

35

Graph Transformations

This topic should appeal to your artistic side (if you have one) — there are lots of beautiful graphs to draw.
Even if you don't, it's still worthwhile as these questions will get you quite a few marks in your exam.

1   The diagram below shows the graph with equation $y = 4^x$.

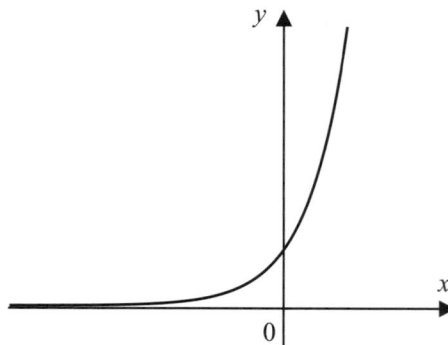

Sketch $y = 4 - 4^x$ on the axes below, labelling any known points of intersection.

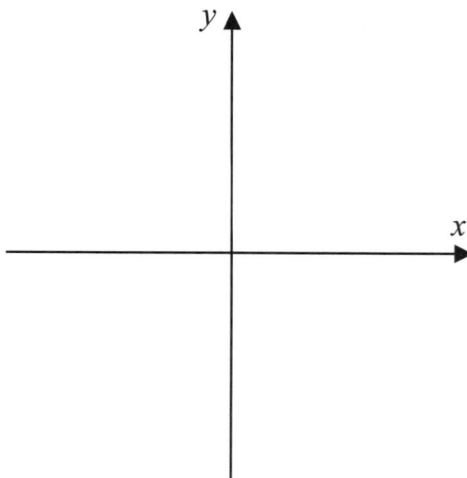

(3 marks)

2   A curve has the equation $y = f(x)$, where $f(x) = (x - 1)^2 (x + 2)^2$.

   a)   Determine the values of x where $f(x - 3)$ touches the x-axis.

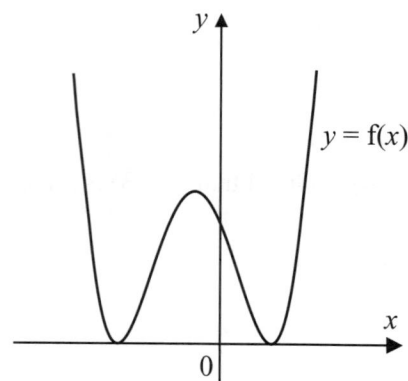

....

(2 marks)

   b)   Given that $g(x) = f(x) + c$, determine the range of values of c for which $g(x)$ has no real roots.

....

(1 mark)

Graph Transformations

3 The diagram on the right shows the graph of $y = f(x)$, with turning points at $(-1, -2)$ and $(3, 2)$.

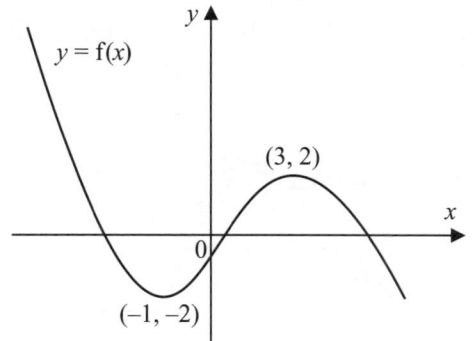

$y = f(x)$

$(3, 2)$

$(-1, -2)$

a) Find the coordinates of the turning points of the graph with equation $y = 3f(x + 2)$.

....

(2 marks)

b) Given that $g(x) = -3f(x + 2)$, sketch the graph of $g(x)$ on the axes below, labelling the coordinates of the turning points of the graph.

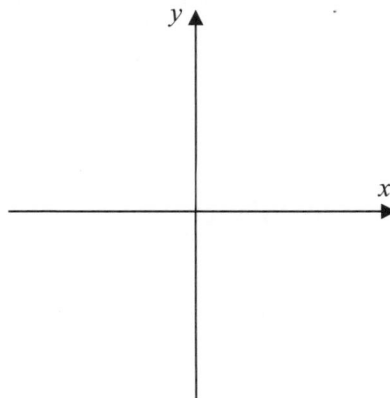

(2 marks)

4 The diagram on the right shows the graphs of $y = f(x)$ and $y = g(x)$, where $g(x) = f(x + a) + b$.

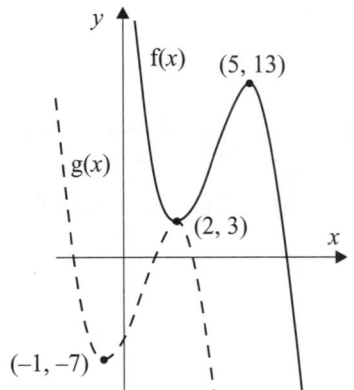

$f(x)$ $(5, 13)$

$g(x)$

$(2, 3)$

$(-1, -7)$

a) Write down the values of a and b.

$a = $ ...., $b = $ ....

(2 marks)

b) P(X, Y) is a point on the graph of $y = f(x)$.
Give the coordinates for the image of point P on the graph of $y = 2g(x + 1)$ in terms of X and Y.

....

(2 marks)

EXAM TIP Remember: $f(x) + a$ and $f(x + a)$ are translations, $af(x)$ and $f(ax)$ are stretches (or squashes), and $-f(x)$ and $f(-x)$ are reflections. Combinations of transformations can look tricky, but they're not so bad if you take them one step at a time — break them down into individual transformations and draw a graph for each stage until you get the final graph.

Score

14

Section One — Algebraic Skills

Exponentials and Logs

Exponentials and logs might seem a bit tricky at first, but once you get used to them and they get used to you, you'll wonder what you ever worried about. Plus, they're really handy for modelling real-life situations.

1   Show that $\dfrac{\ln 54 - \ln 6}{\ln 3} = 2$.

(3 marks)

2   Find the value of $\log_3 243 - \frac{1}{2}\log_3 9$.

....

(3 marks)

3   Given that $\log_a x = \log_a 4 + 3 \log_a 2$, find the value of x.

$x = $ ....

(3 marks)

4   The graph of $f(x) = c^x$ passes through the point $A(3, 64)$.

    a)   Find the value of c.

$c = $ ....

(2 marks)

    b)   Sketch the graph of $f^{-1}(x)$ on the axes below, labelling the image of A and any intersections with the axes.

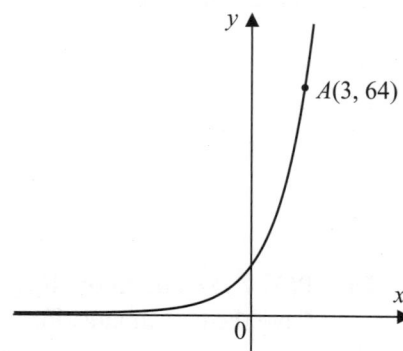

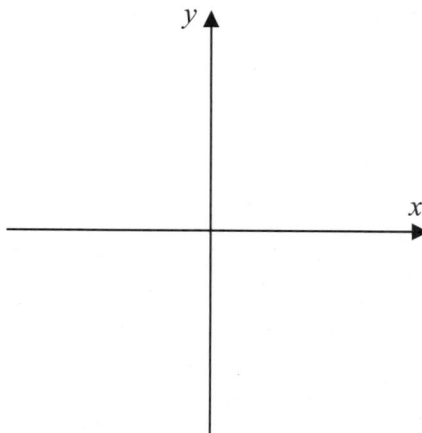

(2 marks)

Exponentials and Logs

5   If $\log_a 54 + \log_a 4 = 3$, find the value of a.

$a =$ ....
(2 marks)

6   Given that $p > 0$, find the value of $\log_p(p^4) + \log_p(\sqrt{p}) - \log_p\left(\dfrac{1}{\sqrt{p}}\right)$.

....
(3 marks)

7   For the positive integers p and q, $\log_4 p - \log_4 q = \dfrac{1}{2}$.

a)   Show that $p = 2q$.

(3 marks)

b)   The values of p and q are such that $\log_2 p + \log_2 q = 7$.
Use this information to find the values of p and q.

$p =$ .... $q =$ ....
(5 marks)

8   For the function $f(x) = 3 \ln x - \ln 3x$, $x > 0$, find the exact value of x when $f(x) = 0$.

$x =$ ....
(3 marks)

Section One — Algebraic Skills

Exponentials and Logs

9 The value, £V, of a large piece of machinery is modelled by $V = pq^t$, where t is the age of the machinery in years and p and q are constants. The line l below plots $\log V$ against t and has a gradient of -0.025.

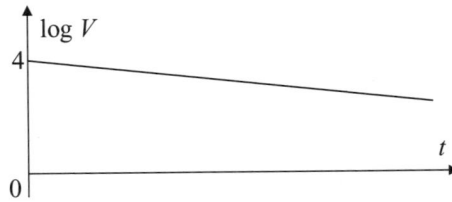

a) Calculate the exact values of p and q.

$p = $ ...., $q = $ ....

(5 marks)

b) Find the value of the machinery when it is 20 years old, to the nearest pound.

£ ....

(2 marks)

10 The spread of a forest fire burning unchecked is modelled by the equation $H = 20e^{bt}$, where H is the area burnt by the fire (in hectares) after t hours.

a) Given that $H = 24$ when $t = 1$, find the value of b to 3 significant figures.

$b = $ ....

(2 marks)

b) Calculate how long it would take the forest fire to burn an area of 500 hectares.

....

(3 marks)

c) The area of the whole forest is 2000 hectares. What percentage of the forest's area will be burning if the fire is left unchecked for 7 hours?

....

(2 marks)

Exponentials and Logs

11 The population of microbes growing in a petri dish, p (1000s), after t days, can be modelled by the formula $p = at^b$, where a and b are constants. The diagram below shows the graph of $\log p$ against $\log t$.

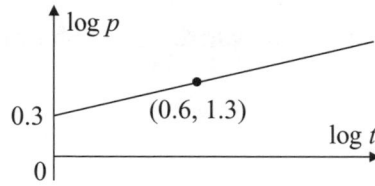

Find the values of a and b to 1 d.p.

$a =$ .... $b =$ ....

(5 marks)

12 The population, P, of an endangered species of bird is modelled over time, t years ($t \geq 0$), by $P = 5700e^{-0.15t}$. The population, Q, of a bird of prey that hunts the endangered species is modelled by $Q = 2100 - 1500e^{-0.15t}$. $t = 0$ represents the beginning of the year 2010, when the bird of prey was first introduced into the country.

a) Predict the year that the population of the endangered species will drop to below 20% of what it was at the start of 2010.

....

(5 marks)

b) Find the year in which the population of the bird of prey is first predicted to exceed the population of the endangered species according to these models.

....

(4 marks)

Score

57

Composite and Inverse Functions

Composite functions are 'functions inside functions', such as f(g(x)), which means "do g(x), then f(the result)".
An inverse function, written like $f^{-1}(x)$, is the 'opposite' of the original function — if f(a) = b, then $f^{-1}(b) = a$.

1   A function f is defined by $f(x) = 4x^2$ for $x \in \mathbb{R}$, where $x \geq 0$. Show that $x = \frac{1}{2}$ when f(f(x)) = 4.

(3 marks)

2   f and g are functions such that $f(x) = 2x^2 + 3$ and $g(x) = \sqrt{2x - 6}$, $x > 3$.

    a)   Find a simplified expression for g(f(x)).

....
(2 marks)

    b)   Evaluate f(g(4)).

....
(2 marks)

3   f(x) and g(x) are defined on suitable domains by $f(x) = x^2 + 2x - 8$ and $g(x) = \frac{1}{2x}$.

    a)   Write down an expression for g(f(x)).

....
(1 mark)

    b)   Find the values of x that cannot be in the domain of g(f(x)).

....
(2 marks)

Composite and Inverse Functions

4   The function f is defined on $\mathbb{R}$, the set of real numbers, by $f(x) = \dfrac{3}{2x + 5}$, where $x \neq -\dfrac{5}{2}$.

   a)   If $g(f(x)) = x$, find $g(x)$.

....

(3 marks)

   b)   Given that $f(2) = \frac{1}{3}$, state the value of $g\!\left(\frac{1}{3}\right)$.

....

(1 mark)

   c)   Evaluate $g(3)$.

....

(2 marks)

5   The functions f and g are defined as follows: $f(x) = 2 \sin x + 1$, $x \in \mathbb{R}$ and $g(x) = \sqrt{3x + 1}$, $x \geq -\dfrac{1}{3}$.

   a)   Find an expression for $g^{-1}(x)$.

....

(3 marks)

   b)   Evaluate $g(f(0))$.

....

(2 marks)

   c)   Find a simplified expression for $g(f(x))$.

....

(2 marks)

EXAM TIP   Remember to always pay attention to the brackets when dealing with composite functions to make sure you do the functions in the right order — f(g(x)) doesn't usually equal g(f(x)). Watch out for questions that say "f(g(x)) = x" — this just means that f and g are inverses of each other, so you can handle them in the same way as in any other inverse functions question.

Score

23

Recurrence Relations

A recurrence relation is a rule for a sequence that says how to get from one number to the next — kind of like a sat-nav that keeps repeating: "× 0.5, + 3, × 0.5, + 3... in ∞ terms, you will have reached your desti- I mean limit".

1 A sequence generates the terms 12, 16, 20, 24, 28...

    a) Write down the recurrence relation for this sequence.

        ....

        (2 marks)

    b) Using your answer to part a), explain whether or not this sequence will approach a limit as $n \to \infty$.

        (1 mark)

2 Calculate the limit of the sequence defined by the recurrence relation $u_{n+1} = \frac{1}{3}u_n + 2$.

        ....

        (2 marks)

3 A sequence is defined by the recurrence relation $u_{n+1} = xu_n - 12$, where $u_0 = 4$.

    a) Find an expression for u_2 in terms of x.

        ....

        (2 marks)

    b) Given that $u_0 = u_2$, find the possible values of x.

        ....

        (3 marks)

Recurrence Relations

4   Find the values of a and b for the sequence $u_{n+1} = au_n + b$, where $u_1 = 6$, $u_2 = 7$ and $u_3 = 8.5$.

$a = $ ...., $b = $ ....

(3 marks)

5   30 rodents were accidentally introduced to an isolated island. Scientists recorded 525 rodents on the island after one year and 921 rodents on the island after two years. The recurrence relation $u_{n+1} = au_n + b$ was used to model the rodent population, where a and b are constants.

a)   Find the values of a and b.

$a = $ ...., $b = $ ....

(4 marks)

b)   Assuming the rodent population continues to increase according to this model, calculate the limit that the population will eventually reach.

....

(2 marks)

6   In 2010, two councils started a scheme to increase the percentage of household waste recycled in their towns. Town A's recycling rate can be modelled by $u_{n+1} = 0.45u_n + 27\%$, $u_0 = 37\%$ and Town B's recycling rate can be modelled by $v_{n+1} = 0.71v_n + 15\%$, $v_0 = 40\%$, where n is the number of years after 2010.

a)   Show that Town A had a higher recycling rate than Town B in 2011.

(3 marks)

b)   The councils set a target to reach a recycling rate of 50%. Determine whether either town will ever reach this target.

(4 marks)

EXAM TIP The good news is that, if you know how to answer all these questions, you should be able to tackle anything they throw at you in the exam. The bad news is that some of the questions will have contexts that can make things confusing. The best strategy is to pick out the key info and try to turn it into a standard recurrence relation question that you know how to handle.

Score

26

Trig Graphs and Transformations

Trig graph transformations are the hottest new thing — transformations of polynomials are so last section...
Actually, they're pretty similar, but it's still important that you practice the trig ones, so no slacking off.

1   The graph of the function $x(t) = a \sin bt + c$ for $0 \leq t \leq \pi$ is shown on the diagram below.
Find the values of a, b and c.

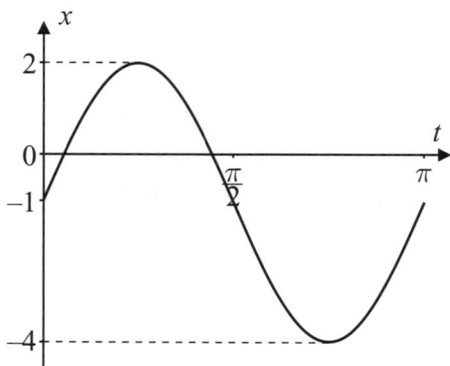

$a =$ ...., $b =$ ...., $c =$ ....

(3 marks)

2   The function f(x) is defined as f(x) = $8 \cos 6x$. Find the period of f(x).

....

(2 marks)

3   The function f(x) = $3 - 2 \cos x$ is defined on the interval $0 \leq x \leq 2\pi$.

   a)   Find the value of x at which the maximum of f(x) occurs.

....

(1 mark)

   b)   The function g(x) is defined as g(x) = $2\text{f}\left(x - \dfrac{\pi}{6}\right)$. Find the minimum value of g(x).

....

(2 marks)

Trig Graphs and Transformations

4   The diagram below shows the graph of f(x) = sin x.

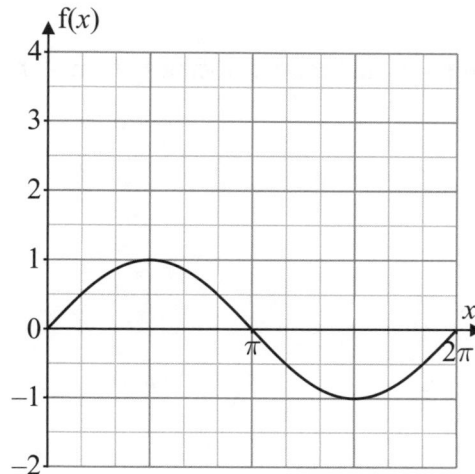

a)   Sketch the graph of g(x) = 2 + f(3x) on the axes above.

(2 marks)

b)   Write down the period of g(x).

....

(1 mark)

5   The function f(t) is defined as f(t) = A + B cos t,
where A and B are positive constants and t is in radians.

   a)   f(t) has a minimum of 7 and a maximum of 17. Find the values of A and B.

A = ... , B = ...
(3 marks)

   b)   The function g(t) is defined as g(t) = A + B cos(Ct + D). Find the values of C and D
   such that g(t) has a maximum at (0, 17) and a minimum at (6, 7).

C = ... , D = ...
(2 marks)

EXAM TIP

Remember that cos and sin can only take values between 1 and –1, so if a graph is doing
something different, it's definitely been transformed in some way — it's your job to figure out
how. It's worth knowing the values of x (in degrees and in radians) where sin and cos are
at their maximums and minimums, but you can always do a quick sketch if you're unsure.

Score

16

Solving Trig Equations

As trig graphs repeat the same shape over and over again, trig equations can have infinitely many solutions. Luckily, you'll only have to solve them for a particular interval, otherwise you would be here for a while...

1 Given that $\cos\theta = \frac{5}{6}$ and $0 \le \theta \le \frac{\pi}{2}$, find the exact values of $\sin\theta$ and $\tan\theta$.

$\sin\theta = $ .... , $\tan\theta = $ ....

(3 marks)

2 The function $f(\theta) = \tan^2\theta + \frac{\tan\theta}{\cos\theta}$ is defined for $0 \le \theta \le 2\pi$, $\theta \ne \frac{\pi}{2}, \frac{3\pi}{2}$.

a) Show that the equation $\tan^2\theta + \frac{\tan\theta}{\cos\theta} = 1$ can be written in the form $2\sin^2\theta + \sin\theta - 1 = 0$.

(4 marks)

b) Hence find all solutions to the equation $f(\theta) = 1$ in the interval $0 \le \theta \le 2\pi$.

....

(3 marks)

3 Given that $\sin x = \frac{8}{9}$ for the acute angle x, find the exact value of $\tan^2 x$.

....

(3 marks)

Solving Trig Equations

4 Find all the solutions of the equation $4\sin 2x - \cos 2x = 0$ in the interval $0 \le x \le 2\pi$.

....

(3 marks)

5 Find all the values of x, in the interval $0° \le x \le 180°$, for which $7 - 3\cos x = 9\sin^2 x$.

....

(5 marks)

6 Find the first three exact values of t for which $12\sin 3\pi t = 6\sqrt{2}$ for $0 \le t \le 2\pi$.

....

(5 marks)

EXAM TIP — Be very, *very* careful with degrees and radians — you need to use the same angle measurement in your answer as the question does, or you'll lose marks. Your calculator will have a 'degrees' mode and a 'radians' mode so make sure you know how to switch from one to the other, and double check — no, triple check — which mode your calculator is in before starting a question.

Score

26

Section Two — Trigonometric Skills

Addition and Double Angle Formulas

As if regular trig wasn't enough, someone's gone and doubled the fun. The double angle and addition formulas for cos and sin are given in the exam, but they won't help if you don't know what they mean or how to use them.

1 The right-angled triangle on the right has an acute angle x such that $\sin x = \frac{2}{3}$ and $\cos x = \frac{\sqrt{5}}{3}$.

a) Find the exact value of $\sin 2x$.

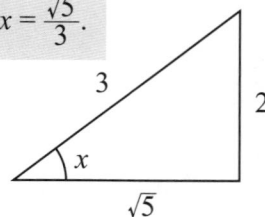

....

(2 marks)

b) Find the exact value of $\cos 2x$.

....

(2 marks)

2 Using a double angle formula, show that $\dfrac{1 + \cos x}{2} \equiv \cos^2 \dfrac{x}{2}$.

Write cos x as cos $2\left(\frac{x}{2}\right)$.

(2 marks)

3 The expression $\cos\left(x - \dfrac{\pi}{4}\right)$ can be written in the form $\dfrac{\sqrt{a}}{b}(\cos x + \sin x)$, where a and b are integers.

a) Determine the values of a and b.

$a =$ ...., $b =$ ....

(2 marks)

b) Hence, or otherwise, find the exact value of $\cos\left(\dfrac{5\pi}{12}\right)$ in the form $\dfrac{\sqrt{c} - \sqrt{d}}{e}$, where c, d and e are integers.

....

(3 marks)

Addition and Double Angle Formulas

4 Solve the equation $\cos 2x + 7\cos x = -4$ for $0° \leq x \leq 360°$.

....
(5 marks)

5 Triangle ABC is shown in the diagram on the right, where angles BAC = x and DAC = y.

Show that $\sin(x - y) = \dfrac{8\sqrt{41}}{205}$.

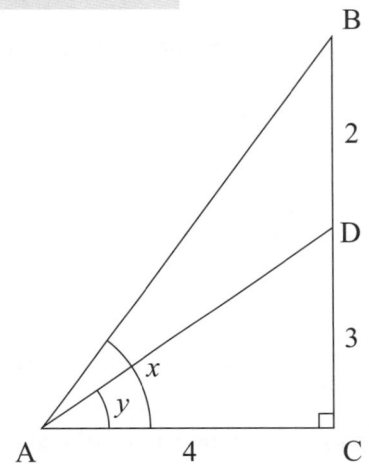

(5 marks)

6 Solve the equation $3\sin 2\theta \tan \theta = 5$ for $0 \leq \theta \leq 2\pi$, giving your answers to 3 significant figures.

....
(5 marks)

EXAM TIP Watch out for those sneaky ± and ∓ signs in the cos addition formula — the sign you use in the long expanded bit is the opposite of the one you use in the bracket. You can use the double angle formula for any multiple of an angle, not just 2x, as long as you change the numbers all the way through the formula — for example, sin 10x = 2 sin 5x cos 5x.

Score

26

The Wave Function

Just in case you weren't convinced from the last two pages about how great addition formulas are, here's another handy use for them — rearranging expressions with both sin and cos into an expression with just one of them.

1   The function $P(\theta)$ is defined as $P(\theta) = 7\sqrt{2}\sin\theta + \sqrt{2}\cos\theta$.

a)   Express $P(\theta)$ in the form $k\cos(\theta - \alpha)$, where $k > 0$ and $0 \leq \alpha \leq 2\pi$.

....

(4 marks)

b)   Hence find the maximum and minimum values of $P(\theta)$.

Maximum = ..., Minimum = ...

(2 marks)

2   The expression $3\cos 2x - 2\sin 2x$ can be written in the form $k\sin(2x + \alpha)$, where $k > 0$ and $0 \leq \alpha \leq 2\pi$.

a)   Find the values of k and α.

$k = $ ...., $\alpha = $ ....

(4 marks)

b)   The diagram on the right shows part of a sketch of the graph of $y = 3\cos 2x - 2\sin 2x$ and the line $y = 1$.

Find the x-coordinates of the points A and B, where the line and the curve intersect.

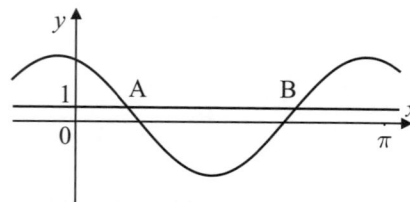

....

(4 marks)

The Wave Function

3   A buoy is floating in a harbour. The height of the buoy above the sea bed, in metres, is modelled by the function $h(t) = 14 + 3\sin t + 5\cos t$, where t is time in minutes. By writing $h(t)$ in the form $14 + k\cos(t + \alpha)$, where $k > 0$ and $0° < \alpha < 360°$, find the maximum height of the buoy and the first time that it reaches that maximum (to 1 d.p.).

Maximum height = ... m , Time = ... mins

(7 marks)

4   A garden sprinkler system is set up to water a flower bed. The distance, d feet, the water sprays at time θ minutes can be modelled by the function $d(\theta) = \sqrt{2}\cos\theta - 3\sin\theta$, where distance to the left of the sprinkler is modelled as negative and distance to the right of the sprinkler is modelled as positive.

a)   Write $\sqrt{2}\cos\theta - 3\sin\theta$ in the form $k\sin(\theta - \alpha)$, where $k > 0$ and $0 \le \alpha \le 2\pi$.

....

(4 marks)

b)   Find the first two times that the water sprays 3 feet to the right of the sprinkler after being switched on. Give your answers in minutes and seconds, to the nearest second.

....

(5 marks)

> **EXAM TIP**
> Pay attention to what range your angle should be in — usually it's between 0° and 360° (or 0 and 2π if it's in radians). Use an ASTC (or CAST) diagram to work out which quadrant the angle should be in, based on whether sin, cos or tan are positive or negative. Otherwise, you'll be throwing away marks after doing all that hard work, and that's the last thing you want.

Score

30

Linear Coordinate Geometry

This topic is all about lines. Unfortunately, lines can get quite complicated when it comes to exam questions. But remember all the important gradient rules and the special properties of lines in triangles and you'll be fine.

1 Find the equation of the line passing through the point (–4, 7), which is parallel to $y = -\frac{1}{2}x + 3$.

....
(2 marks)

2 The points A (3, 7), B (–4, 3) and C (5, –3) form a triangle.

a) Find the equation of line *l*, the altitude from A.

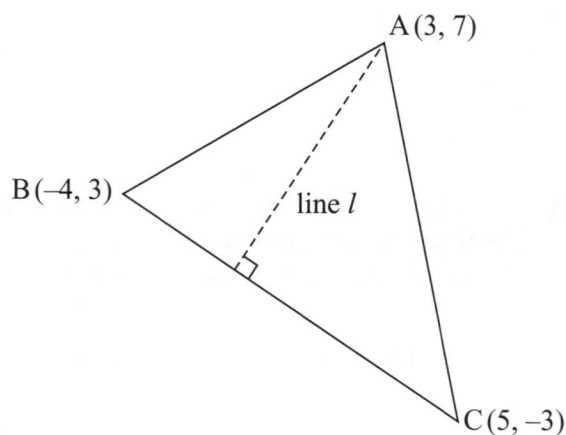

....
(3 marks)

b) Hence, given that the altitude from B has the equation $5y - x = 19$, find the coordinates of the orthocentre of the triangle ABC.

The orthocentre of the triangle is the point where the three altitudes meet.

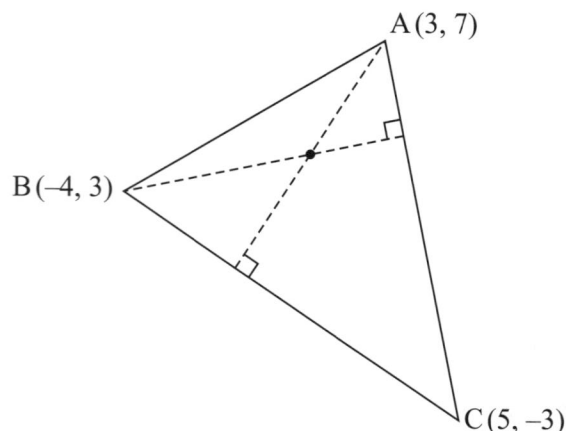

....
(2 marks)

Linear Coordinate Geometry

3 The points A(–2, 4), B(–1, 0) and C(1, 3) form a triangle. Find the equation of line *l*, the median from A.

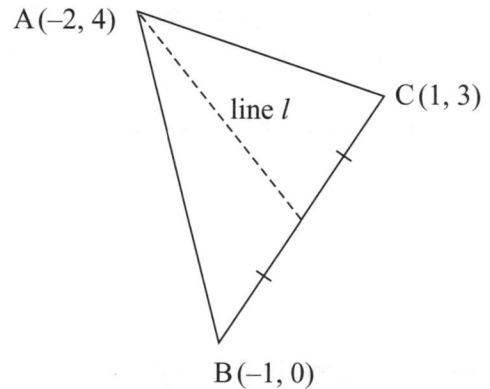

A(–2, 4)

line *l*

C(1, 3)

B(–1, 0)

....

(3 marks)

4 The line *l* has equation $y + 2x - 5 = 0$. Point A lies on *l* and has coordinates (1, k).

a) Find the equation of the line that is perpendicular to *l* and passes through point A.

....

(3 marks)

b) Find the angle *p*, in degrees, that line *l* makes with the positive direction of the *x*-axis.

p = ...°

(2 marks)

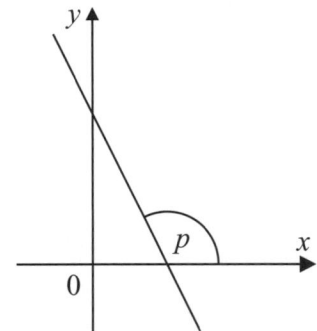

5 Show that the points D(–3, –2), E(–1, 1) and F(5, 10) are collinear.

(3 marks)

Linear Coordinate Geometry

6 The point A lies at the intersection of the lines l_1 and l_2, where the equation of l_1 is $x - y + 1 = 0$ and the equation of l_2 is $2x + y + 8 = 0$.

    a) Find the coordinates of point A.

....
(2 marks)

    b) The points B and C have coordinates $(6, -4)$ and $(-1, 2)$ respectively, and D is the midpoint of AC. Find the equation of the line which passes through B and D.

....
(3 marks)

    c) Show that angle ADB is a right angle.

(2 marks)

7 ABC is a triangle with circumcentre $(3, 2.5)$. The midpoints of AC and BC are $(2, 3)$ and $(4, 3)$, respectively.

    a) Find the coordinates of point C.

> The circumcentre of a triangle is the point where the perpendicular bisectors of the sides meet.

....
(6 marks)

    b) Given that the midpoint of AB is $(3, 1)$, find the equation of AB.

....
(2 marks)

EXAM TIP

If you're not given a diagram in the question, it's a good idea to quickly sketch one. It'll give you a better idea of what the question is asking for, and if you can picture where points are, it's often much easier to see which rules or properties of triangles you can use to reach your answer. Don't worry about it being too accurate — as long as it gives you the general idea.

Score

33

Circle Geometry

You might love circles, or you might hate them — either way you're bound to encounter them in the exam. So give yourself a quick reminder about the properties of circles and get ready to show them that you're not afraid.

1   Write the equation of the circle with radius $\sqrt{5}$ and centre $(3, -2)$.

....
(1 mark)

2   A circle has the equation $x^2 + y^2 - 4x + 1 = 0$.

     a)   Find the coordinates of the centre of the circle.

....
(1 mark)

     b)   Hence, or otherwise, find the radius of the circle.

....
(1 mark)

3   The points $A(2, 1)$ and $B(0, -5)$ lie on a circle, where the line AB is a diameter of the circle.

     a)   Find the centre and radius of the circle.

centre = ..., radius = ...
(2 marks)

     b)   Show that the point $(4, -1)$ also lies on the circle.

(2 marks)

     c)   Find the equation of the tangent to the circle at point A.

....
(3 marks)

Circle Geometry

4 The diagram shows a circle with centre P. The line AB is a chord with midpoint M.

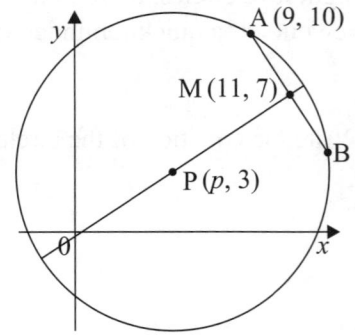

a) Show that $p = 5$.

(3 marks)

b) Find the equation of the circle.

....

(2 marks)

5 The larger of two concentric circles, with centre $(-5, -4)$, has a chord that touches it at A$(-3, 6)$ and B$(b, -6)$, where b is negative. This chord forms a tangent to the smaller circle, as shown in the diagram below.

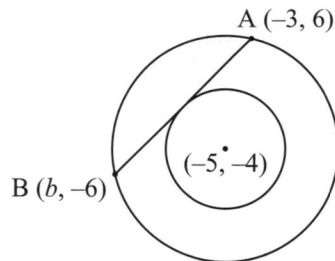

a) Show that $b = -15$.

(4 marks)

b) Find the equation of the smaller circle.

....

(3 marks)

Circle Geometry

6 The circle C has the equation $(x + 2)^2 + (y - 1)^2 = 40$.

a) Show that the point A$(4, 3)$ lies on the circle C.

(1 mark)

b) Find the equation of the tangent to C at the point A.

....

(3 marks)

7 Two circles, C_1 and C_2, touch externally. C_1 has equation $x^2 + y^2 + 6x - 4y = 23$ and C_2 has a radius of $r_2 = 4$ and a centre of $(k, -6)$, where k is positive.

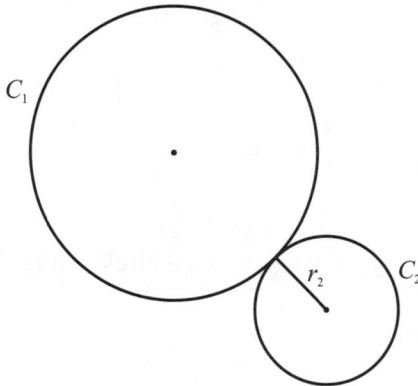

a) Find the value of k.

$k =$ ....

(6 marks)

b) Write down the equation of C_2.

....

(1 mark)

Section Three — Geometric Skills

Circle Geometry

8 Three circles, C_1, C_2 and C_3 all touch at the point A and have radii in the ratio $1:2:5$. The centres of all three circles have the same y-coordinates and C_1 has equation $x^2 + y^2 + 10x - 6y + 25 = 0$.

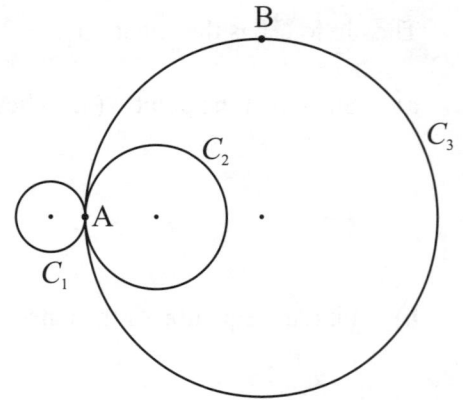

a) Find the equation of C_2.

....

(4 marks)

b) Find the coordinates of point B, the highest point on C_3.

....

(3 marks)

9 The circle C_1 has equation $(x - 4)^2 + (y - 3)^2 = 8$ and the circle C_2 has equation $x^2 + y^2 - 2x + 4y - 3 = 0$.

Show that the circles do not intersect.

(7 marks)

	Score
EXAM TIP If you know that two circles are touching, then you'll probably need to work out the distance, d, between their centres — 'touching externally' means d = the sum of their radii, and 'touching internally' means d = the difference between their radii. If you're not given a diagram in the question, then your first step should be to draw one (and to silently curse those evil examiners).	47

Solving Geometrical Problems

Finding intersections sounds hard, but it's actually just simultaneous equations — solving the equations of two lines or curves simultaneously tells you where they meet. If only all my problems could be solved simultaneously...

1 A circle has the equation $x^2 + y^2 = 5$ and a line has the equation $x - 3y + 5 = 0$.

    a) Show that, where the circle and the line intersect, $y^2 - 3y + 2 = 0$.

(2 marks)

    b) Hence, or otherwise, find the points of intersection of the circle and the line.

....

(3 marks)

2 Find the points of intersection between $y = x^3 + x^2 - 3x + 4$ and $y = 3x + 4$.

....

(5 marks)

3 Determine the points of intersection between the line $x + y = 4$ and the circle $(x + 2)^2 + (y - 8)^2 = 52$.

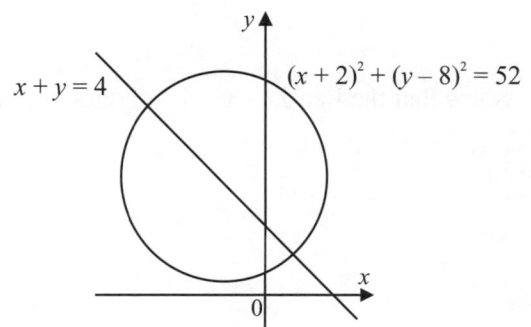

....

(5 marks)

Solving Geometrical Problems

4   Show that $2y - x + 2 = 0$ is a tangent to the circle $(x + 7)^2 + (y + 2)^2 = 5$
and find the coordinates of their point of intersection.

....

(5 marks)

5   Find the coordinates of the points where the curves
$y = x^3 + 5x^2 - 2x - 15$ and $y = x^3 - x^2 + 4x - 3$ intersect.

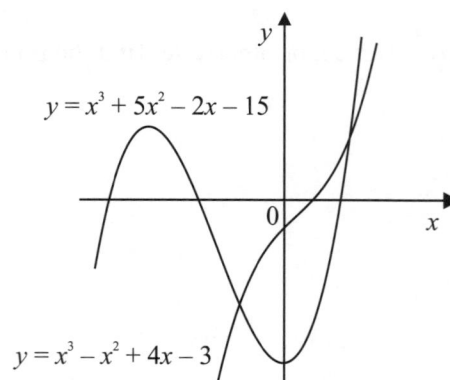

$y = x^3 + 5x^2 - 2x - 15$

$y = x^3 - x^2 + 4x - 3$

....

(5 marks)

6   Show that the line $2x - y - 1 = 0$ does not intersect the circle $(x + 3)^2 + (y - 3)^2 = 15$.

(4 marks)

EXAM TIP These types of questions can seem intimidating at first, but after a bit of practice you'll find that a lot of them are quite similar. For example, as they all involve intersections, you know that at some point you'll need to set two things equal to each other or substitute one equation into another. Then it's just a case of either showing that they do or don't intersect by rearranging.

Score

29

Vectors

Aaah, good old dependable vectors. They've got a certain magnitude about them, and they always have a clear direction. Much like an inspirational leader. It's kind of how I see myself after I've brought about the revolution.

1 Vectors **a** and **b** are given by $\begin{pmatrix} 2 \\ 3 \\ -2 \end{pmatrix}$ and $\begin{pmatrix} 1 \\ 4 \\ 1 \end{pmatrix}$ respectively.

  a) Find the vector $3\mathbf{a} - 2\mathbf{b}$.

....
(1 mark)

  b) Hence, calculate $|3\mathbf{a} - 2\mathbf{b}|$.

....
(2 marks)

2 The vector **a** is given by $\begin{pmatrix} 4 \\ -4 \\ 7 \end{pmatrix}$. Find the unit vector in the direction of **a**.

....
(3 marks)

3 The diagram on the right shows a 3D prism, where $\overrightarrow{DC} = \mathbf{r}$, $\overrightarrow{DA} = \mathbf{s}$, $\overrightarrow{DE} = \mathbf{t}$.

  a) Given that M is the midpoint of AC, find $\overrightarrow{AM}$ in terms of **r** and **s**.

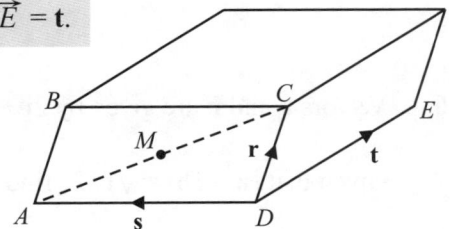

$\overrightarrow{AM}$ = ...
(2 marks)

  b) Find $\overrightarrow{ME}$ in terms of **r**, **s** and **t**.

$\overrightarrow{ME}$ = ...
(2 marks)

Section Three — Geometric Skills

Vectors

4 Points A, B and C have position vectors given by $\overrightarrow{OA} = -5\mathbf{i} + 3\mathbf{j} + 2\mathbf{k}$, $\overrightarrow{OB} = \mathbf{i} - 7\mathbf{j} + 10\mathbf{k}$ and $\overrightarrow{OC} = 3\mathbf{i} + 2\mathbf{j} + 3\mathbf{k}$.

Given that M is the midpoint of AB, find the value of k such that $|\overrightarrow{AB}| = k|\overrightarrow{CM}|$.

$k = \dots\dots\dots\dots\dots\dots\dots\dots$

(5 marks)

5 Given that $\overrightarrow{OA} = -2\mathbf{i} + 4\mathbf{j} - 5\mathbf{k}$, $\overrightarrow{OB} = 14\mathbf{i} + 12\mathbf{j} - 9\mathbf{k}$ and $\overrightarrow{OC} = 2\mathbf{i} + \mu\mathbf{j} + \lambda\mathbf{k}$, find the values of μ and λ such that points A, B and C are collinear.

A, B and C will be collinear if $\overrightarrow{AB}$ is parallel to $\overrightarrow{AC}$

$\mu = \dots\dots\dots\dots, \quad \lambda = \dots\dots\dots\dots$

(5 marks)

6 Vectors **a** and **b** are given by $2\mathbf{i} + \mathbf{j} + 3\mathbf{k}$ and $4\mathbf{i} + m\mathbf{j} - \mathbf{k}$ respectively, where m is an integer.

Given that $|\mathbf{a} - 3\mathbf{b}| = \sqrt{152}$, find the value of m.

$m = \dots\dots\dots\dots\dots\dots\dots\dots$

(4 marks)

Vectors

7   Points A, B and C have position vectors given by $\overrightarrow{OA} = \begin{pmatrix} 1 \\ -3 \\ 2 \end{pmatrix}$, $\overrightarrow{OB} = \begin{pmatrix} 4 \\ -9 \\ 8 \end{pmatrix}$ and $\overrightarrow{OC} = \begin{pmatrix} -3 \\ 9 \\ -4 \end{pmatrix}$.

Point D divides the line AB in the ratio $2:1$.

Point E is positioned such that $\overrightarrow{OD} = -\frac{1}{2}\overrightarrow{CE}$.

Find the position vector of point E.

....

(4 marks)

8   The diagram on the right shows a cube.

Vectors **a** and **b** are given by $-7\mathbf{i} - 4\mathbf{j} + 4\mathbf{k}$ and $m\mathbf{i} + 8\mathbf{j} + \mathbf{k}$ respectively, point O is the origin and point R has position vector $-3\mathbf{i} - 3\mathbf{j} + 12\mathbf{k}$.

a)   Given that m is negative, find the value of m.

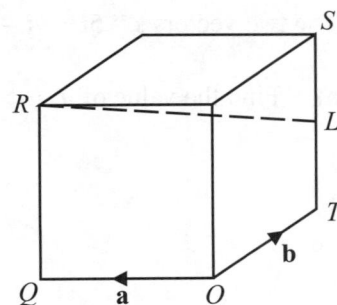

$m =$ ....

(2 marks)

b)   Given that L is the midpoint of TS, find the exact value of $|\overrightarrow{RL}|$.

....

(4 marks)

EXAM TIP   Make sure you're happy with finding the magnitude of a vector as it'll be useful in lots of different questions. Some questions can involve complex shapes, so start by looking at the diagram (or drawing one if you need to) and adding any other information you can get from the question. P.S. Thanks for finding ME back there in question 3b, I'd been lost for ages...

Score

34

Section Three — Geometric Skills

The Scalar Product

The scalar product is a way of multiplying vectors together. You can use it to work out the angle between any two vectors. Remember that the scalar product is shown using a dot, so if you see one, it's ~~dot~~ not a typo.

1 The vectors **a** and **b** are given by $\mathbf{a} = -3\mathbf{i} + 5\mathbf{j} - 2\mathbf{k}$ and $\mathbf{b} = 4\mathbf{i} + 3\mathbf{j} - 6\mathbf{k}$.

    a) Evaluate **a.b**

....

(1 mark)

    b) Hence, calculate the size of the acute angle between **a** and **b** in degrees to 1 decimal place.

....°

(4 marks)

2 The two vectors $\mathbf{s} = 5\mathbf{i} + u\mathbf{j} - 8\mathbf{k}$ and $\mathbf{t} = 6\mathbf{i} + 2\mathbf{j} + \mathbf{k}$ are perpendicular.

    a) Find the value of u.

$u = $ ....

(2 marks)

    b) Hence, evaluate the scalar product **s.(s + t)**

....

(3 marks)

3 The diagram on the right shows triangle PQR, where $\overrightarrow{PQ} = \begin{pmatrix} 2 \\ -9 \\ 3 \end{pmatrix}$ and $\overrightarrow{QR} = \begin{pmatrix} 14 \\ 6 \\ 7 \end{pmatrix}$.

Find the angle QPR, giving your answer to 1 decimal place.

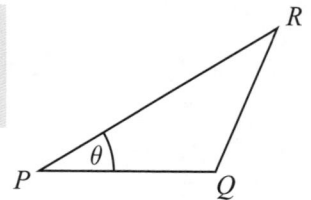

....°

(6 marks)

The Scalar Product

4 The diagram on the right shows the triangular-based pyramid A,OBC, where O is the

origin and points A, B and C have position vectors $\begin{pmatrix} 3 \\ -2 \\ 6 \end{pmatrix}$, $\begin{pmatrix} 6 \\ 0 \\ 0 \end{pmatrix}$ and $\begin{pmatrix} 5 \\ -6 \\ 0 \end{pmatrix}$ respectively.

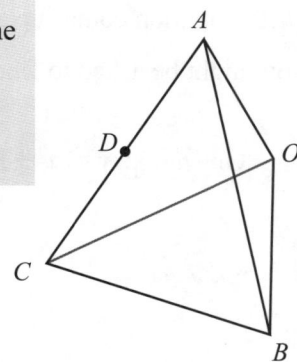

    a) Find the position vector of point D, the midpoint of AC.

....

(2 marks)

    b) Hence, calculate the size of angle DOB, giving your answer in degrees to 3 s.f.

°

....

(5 marks)

5 The diagram on the right shows a square $ABCD$, and an equilateral triangle CDE.

Given that $\overrightarrow{AB}.\overrightarrow{AE} = 32$, calculate $|\mathbf{t}|$.

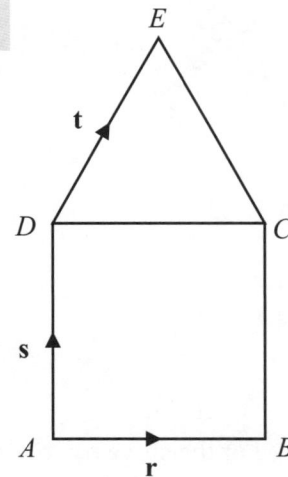

....

(5 marks)

Score

28

Differentiation

Differentiation comes up all over the place in mathematics, so it's worth making sure you're confident with it. You might be asked to find $\dfrac{dy}{dx}$ for an equation y, or $f'(x)$ for a function $f(x)$, but they're both the same thing.

1   Given that $y = x^7 + \dfrac{2}{x^3}$, find $\dfrac{dy}{dx}$.

....

(3 marks)

2   For the curve given by the equation $y = 3\sin 2x$, find $\dfrac{dy}{dx}$.

....

(2 marks)

3   A curve has equation $y = 3x + 4 + x^4$.

         The point A (2, 26) lies on the curve. Find the gradient of the curve at point A.

....

(3 marks)

4   For the curve $y = x^5 - 4x^3 + \dfrac{1}{x}$, show that the tangents to the curve at $x = a$ and $x = -a$ are parallel for all values of a.

(4 marks)

Differentiation

5 A function f(x) is defined on the set of real numbers by f(x) = $2\sin^3 x$. Find the value of $f'\left(\dfrac{\pi}{3}\right)$.

....

(3 marks)

6 The curve C is given by the equation $y = 2x^3 - 10x^2 - 4\sqrt{x} + 12$.

Find the gradient of the tangent to the curve at the point where $x = 4$.

....

(4 marks)

7 Find $\dfrac{dy}{dx}$ at the given point for each of the following:

a) $y = (6x + 4)^3$, (−1, −8)

....

(3 marks)

b) $y = \dfrac{1}{\sqrt{2x - x^2}}$, (1, 1)

....

(4 marks)

8 Find the equation of the tangent to the curve $y = (3 - 2x)^7$ at $x = 2$.

....

(5 marks)

Section Four — Calculus Skills

Differentiation

9 Find the exact gradient of the curve $y = \cos 2x + \sin\left(x + \frac{\pi}{3}\right)$ at the point $\left(-\frac{\pi}{3}, -\frac{1}{2}\right)$.

....

(3 marks)

10 A curve C is given by the equation $y = x^3 - 7x^2 + 9x + 12$.

    a) Find the equation of the tangent to the curve at the point where it crosses the y-axis.

....

(4 marks)

    b) Determine the coordinates of the point where the tangent meets the curve C again.

....

(4 marks)

11 A function is given by $f(x) = \dfrac{x^3 - 5x^2 - 4x}{x\sqrt{x}}$ for $x > 0$.

Find the equation of the tangent to $f(x)$ at the point $(4, -4)$.

....

(5 marks)

EXAM TIP It's worth getting lots of practice at differentiating because there is absolutely no escaping it. Remember that the gradient of a curve at a point is equal to the gradient of the tangent at that point. If you're differentiating a polynomial, write all of the terms as powers of x — it's much easier to see what's going on without a load of fractions and roots getting in the way.	**Score** **47**

Stationary Points

Stationary points occur where the gradient of a curve is 0. That's a pretty important starting point for these next few pages, so make sure you're 100% happy with that before you go any further. All good? Brilliant, off we go...

1   A function $f(x)$ is defined by $f(x) = x^3 - 5x^2 + 3x + 9$.
   Determine whether $f(x)$ is increasing or decreasing when $x = 4$.

....

(3 marks)

2   Find the range of x values where the function $y = 5 - 3x - x^2$ is decreasing.

....

(3 marks)

3   The graph of the function $f(x)$ is shown below. Use it to sketch the graph of $f'(x)$.

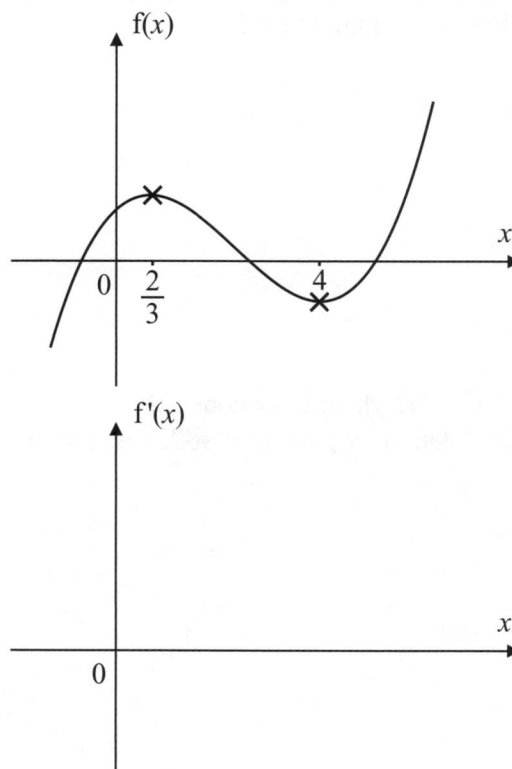

(3 marks)

Stationary Points

4   A function f(x) is given by f(x) = $7 - 4x - x^3$.

a)   Find f'(x).

....

(2 marks)

b)   Hence, or otherwise, show that f(x) is strictly decreasing for all values of x.

(2 marks)

5   The graph of the curve $y = x^4 - ax^3 - 18x^2 + 108x$ has two stationary points, one at $x = 3$ and one at $x = -3$.

a)   Find the value of a.

$a = $ ....

(3 marks)

b)   Determine the nature of the stationary point at $x = 3$.

....

(2 marks)

6   The function f(x) is given by f(x) = $x^3 + kx$, where k is a constant.
Given that the graph of $y = $ f(x) has a stationary point at $x = -2$, find the value of k.

$k = $ ....

(3 marks)

Stationary Points

7 Erin claims that the function $f(x) = 3x^3 + 9x^2 + 25x$ is a strictly increasing function for all values of x.

Show that Erin's claim is correct.

(4 marks)

8 The function $f(x) = 2x^4 + 64x$ has one stationary point.

a) Find the coordinates of the stationary point.

....
(4 marks)

b) Find the range of values of x for which the function is increasing and the range of values of x for which it is decreasing.

Increasing for: ...., decreasing for: ....
(2 marks)

c) Hence sketch the curve $y = f(x)$, showing where it crosses the axes.

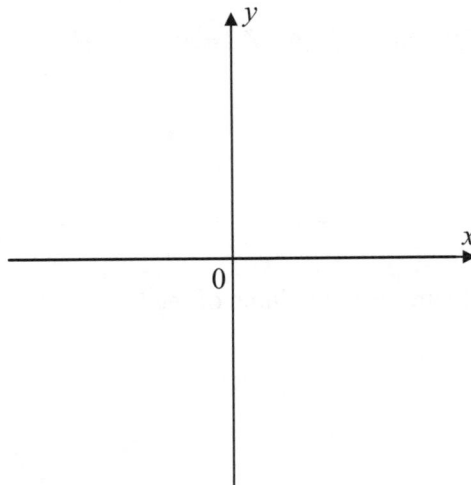

(3 marks)

EXAM TIP Sketching a graph can be difficult, so make sure you find out as much information about the graph as you can first, like knowing what the gradient is doing in different parts of the graph. To check what happens as x gets big, factorise f(x) by taking the highest power of x outside the brackets. Then if you imagine x getting very big, you can see what f(x) will tend towards.

Score

34

Using Differentiation

Differentiation has loads of uses — finding the gradient of a curve barely scratches the surface. It's a useful way to find the maximum or minimum value that an equation can take in a given situation. Fascinating stuff.

1   A ball is catapulted vertically into the air. After t seconds, the height h of the ball, in metres, is given by $h = 30t - 5t^2$. Find the maximum height the ball reaches.

.... m
(5 marks)

2   A curve is defined by the equation $y = 3x^2 - 8x^{\frac{3}{2}}$ for $0 \le x \le 9$.

a)   Find the x-coordinates of the stationary points on the curve.

....
(4 marks)

b)   Determine the greatest and least values of y on this interval.

Greatest value: ...., Least value: ....
(3 marks)

3   A pet food manufacturer designs cylindrical tins of cat food. Each tin is made from 100π cm² of aluminium, and has a radius r cm and height h cm.

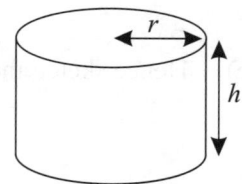

a)   Show that the volume, V, of the tin is given by $V = 50\pi r - \pi r^3$.

(3 marks)

b)   Find the value of r which maximises the volume of the tin.

$r = $ ....
(6 marks)

Using Differentiation

4 An ice cream parlour needs an open-top stainless steel container with a capacity of 40 000 cm³, modelled as a cuboid with sides of length x cm, x cm and y cm, as shown in the diagram.

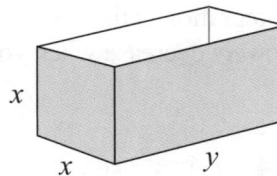

a) Show that the external surface area, A cm², of the container is given by $A = 2x^2 + \dfrac{120\,000}{x}$.

(3 marks)

b) Find the value of x that minimises the surface area of the container.

.... cm
(6 marks)

c) Calculate the minimum area of stainless steel needed to make the container.

.... cm²
(2 marks)

5 The time it takes Paul to bake n cookies is given by $\left(\dfrac{1}{5}n^2 + 50\sqrt{n}\right)$ minutes for $n > 0$. Find the value of n that minimises the amount of time required per cookie.

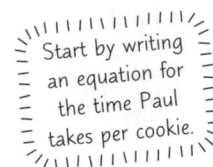

Start by writing an equation for the time Paul takes per cookie.

$n = $ ....
(7 marks)

EXAM TIP These questions can get pretty tricky, so keep a cool head and don't forget it's just the same old differentiation stuff you've already done. If the question asks for a maximum you need to show that the answer you found actually gives the maximum — use a nature table (or find the second derivative) to be on the safe side. (And, well, to get all the marks.)

Score

39

Integration

Remember, integration is the reverse process of differentiation, so powers should <u>in</u>crease when you <u>in</u>tegrate — just like powers <u>de</u>crease when you <u>de</u>fferentiate (okay, it doesn't work quite as well but you get the idea).

1 Find $\int (4x^3 + 6x + 3) \, dx$.

....
(2 marks)

2 Find $\int 4 \sin x \, dx$.

....
(2 marks)

3 A curve that passes through the point (0, 0) has derivative $\dfrac{dy}{dx} = 3x^2 + 6x - 4$.

Find the equation of the curve.

> You're given the derivative of the curve — so integrate to find its equation.

....
(4 marks)

4 The gradient of the curve $y = f(x)$ is given by $\dfrac{dy}{dx} = 6 \cos\left(2\left(x - \dfrac{\pi}{6}\right)\right)$.
The curve passes through $\left(\dfrac{\pi}{6}, 3\right)$. Express y in terms of x.

....
(4 marks)

Integration

5   The curve D has the equation $y = f(x)$, $x > 0$, where $f'(x) = 2x + 3\sqrt{x} + \dfrac{12}{x^2}$.

Given that a point P on curve D has the coordinates $(4, 17)$, find $f(x)$.

....

(5 marks)

6   Find $\displaystyle\int \left(\dfrac{x^2 + 3}{\sqrt{x}} \right) \mathrm{d}x$.

....

(3 marks)

7   The gradient of a curve is given by $\dfrac{\mathrm{d}y}{\mathrm{d}x} = (8x + 1)^{-\frac{1}{2}}$, $x > -\dfrac{1}{8}$.

Given that the curve passes through $(3, 1)$, find y in terms of x.

....

(4 marks)

8   For the function $f(x) = \left(7 - \dfrac{4}{3}x \right)^{-\frac{1}{2}}$:

a)   Find $f'(x)$.

....

(2 marks)

b)   Hence find $\displaystyle\int -2\left(7 - \dfrac{4}{3}x \right)^{-\frac{3}{2}} \mathrm{d}x$.

....

(1 mark)

Integration

9 For a function f on a suitable domain, $f'(x) = 5\sin\left(3x + \frac{\pi}{6}\right)$ and $f\left(\frac{\pi}{2}\right) = \frac{7}{6}$. Find $f(x)$.

(4 marks)

10 Find y in terms of x, given that $\frac{dy}{dx} = \frac{3x^2 - 5\sqrt{x}}{x^4}$ and that $y = -2$ when $x = 1$.

(5 marks)

11 The function $f(x)$ has derivative $f'(x) = 2\cos(4 - 6x)$.

a) Find $f(x)$.

(2 marks)

b) Show that $2\cos(4 - 6x) \equiv 2 - 4\sin^2(2 - 3x)$. Write $4 - 6x$ as $2(2 - 3x)$.

(2 marks)

c) Hence, or otherwise, find $\int (8\sin^2(2 - 3x) - 4)\,dx$.

(2 marks)

EXAM TIP
Knowing your index laws inside out is a big help when it comes to integration — always write out fractions and roots as powers of x before integrating (it'll save a lot of heartache further down the line). The examiners might try to confuse you by sneaking some extra algebra in there or wording the question weirdly, but the basic integration process is the same each time.

Score

42

Definite Integrals

Definite integrals are just like regular old integrals, except for one small difference. Definite integrals have limits — those little numbers by the integral sign that tell you which values of x you need to substitute in at the end.

1   Find the exact value of $\int_1^3 3x^2 - 4x \, dx$.

....
(3 marks)

2   Find the exact value of $\int_{\frac{\pi}{12}}^{\frac{\pi}{8}} \sin 2x \, dx$.

....
(3 marks)

3   Evaluate $\int_0^1 (6x + 1)^{-3} \, dx$.

....
(3 marks)

4   The curve $y = 2\sin 3x$ is shown below. Given that the shaded area is $\frac{4}{3}$ units^2, find $\int_0^\pi 2\sin 3x \, dx$.

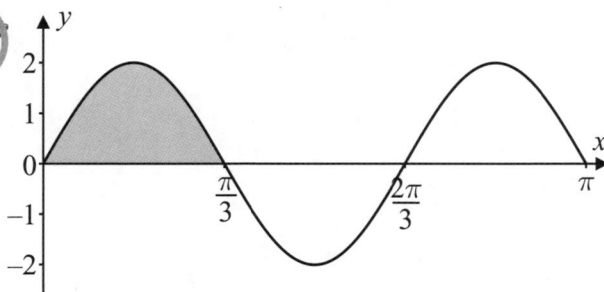

.... units^2
(2 marks)

Definite Integrals

5 Evaluate the integral of the function $y = 5\cos\left(\dfrac{x}{6} - \pi\right)$ between $x = -\pi$ and $x = \pi$.

....

(4 marks)

6 The curve $y = x^3 - 3x^2 - 6x + 8$ is shown below. It crosses the x-axis at $(-2, 0)$, $(1, 0)$ and $(4, 0)$.

Find the area of the shaded region bounded by the curve, the x-axis and the lines $x = -1$ and $x = 3$.

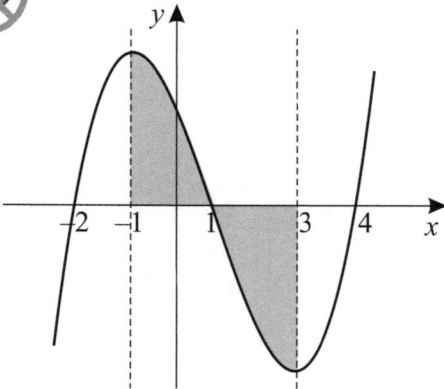

.... units^2

(6 marks)

7 Find the possible values of k that satisfy $\displaystyle\int_{\sqrt{2}}^{2} (8x^3 - 2kx)\,\mathrm{d}x = 2k^2$, where k is a constant.

$k = $ ....

(6 marks)

Definite Integrals

8 The graph below shows the curve C, which has equation $y = \sqrt{x} - \frac{1}{2}x^2 + 1 \ (x \geq 0)$.

Point M lies on the curve and has coordinates $\left(1, \frac{3}{2}\right)$.

Line N has the equation $y = 2x - \frac{1}{2}$ and crosses the curve C at point M.

The shaded region A is bounded by the y-axis, the curve and the line N.

Find the area of A.

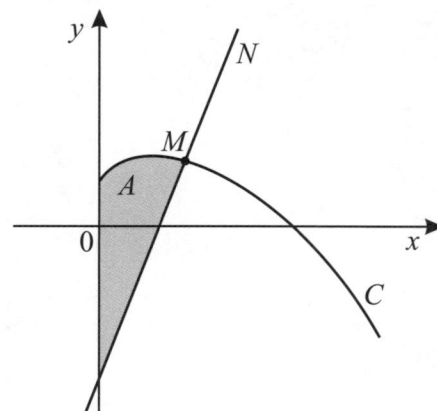

.... units²

(5 marks)

9 A heart shape has been modelled by parts of the four functions shown in the sketch.

- $a(x) = \sin\left(\frac{\pi x}{3}\right) + 4$

- $b(x) = -\sin\left(\frac{\pi x}{3}\right) + 4$

- $c(x) = x - 2$

- $d(x) = -x + 4$

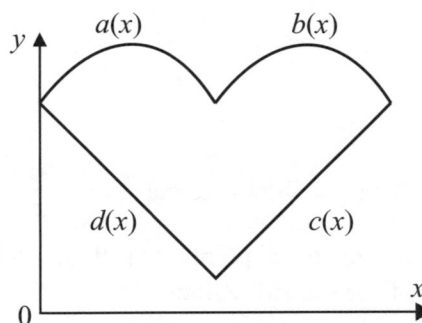

Given that the heart has a vertical line of symmetry, calculate the area of the heart to 3 s.f.

.... units²

(7 marks)

Section Four — Calculus Skills

Definite Integrals

10 The shaded region in the diagram below is bounded by $y = 3x^2 - 6x + 5$ and $y = 14$.

   a) Find the total area of the shaded region.

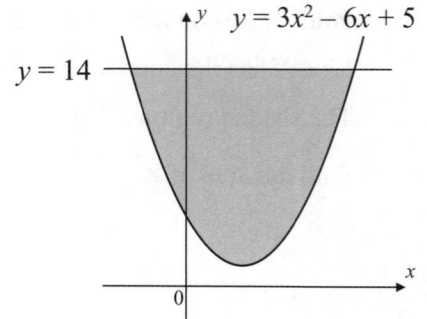

.... units^2
(6 marks)

   b) Determine the ratio of the shaded area on the left of the y-axis to the shaded area on the right.

....
(4 marks)

11 The graph on the right shows the curves $y = \dfrac{9}{x^2}$ and $y = 10 - x^2$ for $x, y > 0$.

The two curves intersect at points A (1, 9) and B (3, 1).
Find the area of the shaded region.

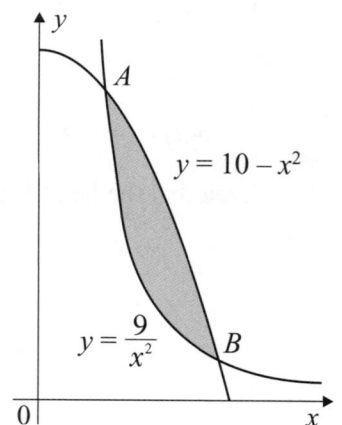

.... units^2
(6 marks)

EXAM TIP
The best thing about definite integrals is that there's no need to worry about that pesky little '+ C'. Finding the area under a curve can be tricky — watch out for curves that go above and below the x-axis and integrate them in separate chunks so that the positive and negative areas don't cancel each other out. If you're not sure what the curve looks like, do a quick sketch first.

Score

55

Rates of Change

'Rates of change' sounds complicated, but it's really just a different way of thinking about differentiation and integration — and since you'll be an expert at both by now, this stuff shouldn't give you too much trouble.

1   Find the rate of change of the function $f(x) = \dfrac{1}{(3x+1)^2}$ when $x = -1$.

....
(3 marks)

2   A particle moves along a path described by the equation $x = 2t^3 - 4t^2$, $t \geq 0$, where t is the time in seconds and x is the displacement in metres.

    a)   Given that velocity is the rate of change of displacement, find a formula for the velocity of the particle after t seconds.

....
(2 marks)

    b)   Find t and x when the speed is 30 m/s.

$t =$ .... s, $x =$ .... m
(3 marks)

    c)   Given that acceleration is the rate of change of velocity, find the acceleration of the particle after 4 seconds.

.... m/s^2
(3 marks)

3   Calculate the rate of change of $P(x) = 4\cos^4 x$ at $x = \dfrac{\pi}{4}$.

....
(3 marks)

Section Four — Calculus Skills

Rates of Change

4 The rate of change of the volume of a siren V, in decibels (dB), is modelled by $\frac{dV}{dt} = 64\sin\left(4t + \frac{\pi}{3}\right)$, where t is the time in seconds, $0 \leq t \leq 10$.

a) Show that the volume of the siren is increasing at $t = \frac{\pi}{8}$ seconds.

(2 marks)

b) Given that the volume of the siren is 120 dB at $t = \frac{\pi}{6}$ seconds, find an expression for the volume of the siren in terms of t.

....

(4 marks)

c) Calculate the volume of the siren at $t = \frac{\pi}{2}$ seconds.

.... dB

(2 marks)

5 The rate of change of the height of a flower H, in cm, is modelled by $\frac{dH}{dt} = \frac{3\sqrt{t}}{4} + k$, where t is the time in days, $0 \leq t \leq 10$. The plant starts at $H = 0$ when $t = 0$, and after 4 days, the plant is 10 cm tall.

By finding the value of k, express the height of the flower, H cm, in terms of t only.

....

(6 marks)

EXAM TIP Rate of change questions often come with odd contexts and most of those contexts are about how things change over time. That's why you'll see t as a variable in so many rates of change questions — don't let it throw you off. No matter the letter used in the question, it doesn't change the maths one bit — so go forth and differentiate and integrate to your heart's content.

Score

28

SQA Higher Mathematics

Practice Paper 1 (Non-Calculator)

Duration — 1 hour 30 minutes

Fill in these boxes and read what is printed below.

Full name of centre

Town

Candidate Forename(s)

Candidate Surname

Date of birth

Day	Month	Year

Scottish candidate number

Total marks available for this paper — 70

You may NOT use a calculator.

Answer ALL the questions.

Write your answers in the spaces provided.

You must use **BLUE** or **BLACK** ink to write your answers.

To earn full marks you must show all of your working.

Give units with your answer where required.

You will find the formula list at the back of this book.

Total marks — 70

Attempt ALL questions

1. Evaluate:

   (a) $\log_2 32$

   1

   (b) $\log_4 16 + \log_{16} 4$

   2

2. A circle has centre (1, –3) and the point (–3, 6) is on its circumference. Find the equation of this circle.

   2

3. Express $4x^2 - 8x + 9$ in the form $a(x + b)^2 + c$.

   3

4. Show that $\dfrac{\cos 2x}{\cos x} + \sin x \tan x \equiv \cos x$, where $\cos x \neq 0$.

   2

5. The functions f and g are defined on the set of real numbers by:

$$f(x) = 2\cos(3x - 1) \qquad g(x) = 8x^3 + 3$$

Find an expression for:

(a) $g(f(x))$

2

(b) $g^{-1}(x)$

3

6. The motion of an oscillating object is modelled using the wave function

$$s(t) = 4\sin 2t - 1,$$

where t is measured in seconds. Calculate:

(a) the period of this function.

1

(b) the maximum and minimum values of $s(t)$.

2

7. The quadratic function $h(x)$ is defined as $h(x) = x^2 - (m + 2)x + (6 - m)$, where m is a constant, for all real numbers x.

Find the values of m for which the equation $h(x) = 0$ has equal roots.

4

62

MARKS

8. The graph of $y = x^4 - 6x^2 - 8x + 5$ has a stationary point at $x = -1$.

(a) Find $\dfrac{dy}{dx}$.

2

(b) Find the x-coordinates of any other stationary points on the graph.

3

(c) Hence, or otherwise, determine the nature of the stationary point at $x = -1$.

2

9. Triangle ABC is a right-angled triangle with angle BAC = x.
Point D lies on the line AC such that angle CBD = y.
The lengths of each side are shown on the diagram.

Show that $\sin(x + y) = \dfrac{13\sqrt{10}}{50}$

3

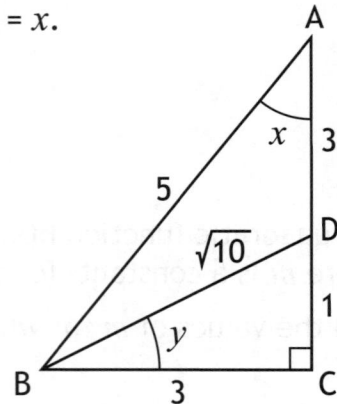

10. The diagram below shows triangle PQR, with vertices at
P(−4, 2), Q(5, 5) and R(8, −4). The line L_1 is the altitude from Q.

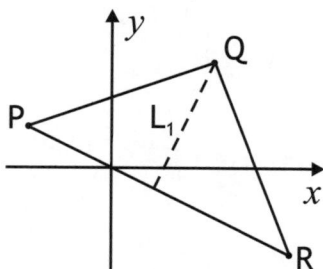

(a) Find the equation of the line L_1.

3

The point S is placed such that the perpendicular bisector of PS passes through the origin and makes an angle of $a°$ with the positive direction of the x-axis, where $0° < a < 90°$ and $\tan a = \frac{1}{3}$.

(b) Work out the equation of the line PS.

3

(c) Find the coordinates of the point where L_1 crosses the perpendicular bisector of PS.

2

11. Find $\int_0^{\frac{\pi}{6}} 2\sin\left(3x - \frac{\pi}{2}\right) dx$.

4

12. Triangle ABC has vertices at A(-7, 4, 3), B(-2, 2, a) and C(0, 1, -1).

Find the value(s) of a for which the vectors $\overrightarrow{AB}$ and $\overrightarrow{BC}$ are perpendicular.

6

13. The diagram on the right shows the graph of the cubic function $y = f(x)$. The graph has turning points at $(0, -3)$ and $(6, 9)$.

On the axes below, sketch the graph of $y = \frac{1}{3}f(2x) - 1$.

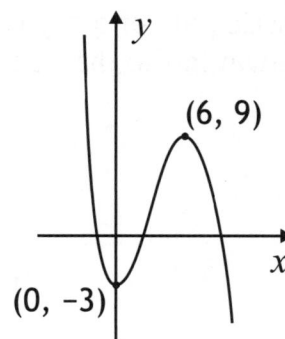

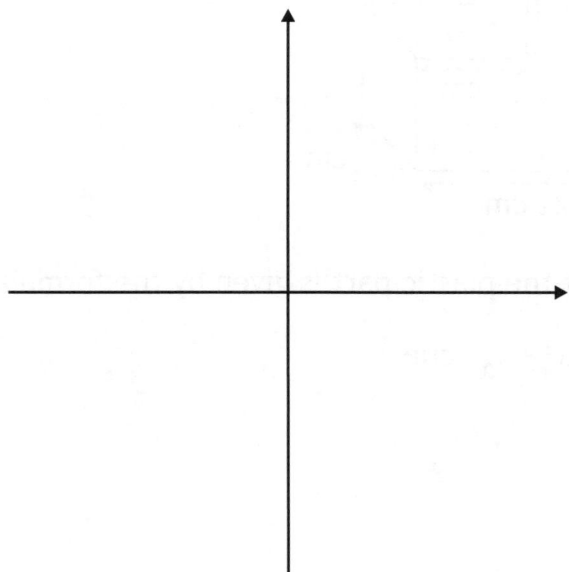

(6, 9)

(0, −3)

3

14. The line $y = ax + 1$ is a tangent to the circle $(x - 3)^2 + (y - 2)^2 = 5$. Find the two possible values of a.

6

66

15. A plastic part of a toy is an L-shaped prism with a volume of 36 cm³, as shown in the diagram below.

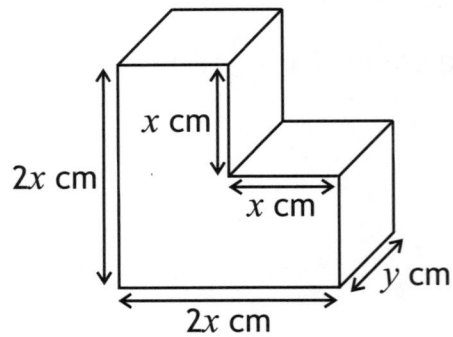

(a) Show that the surface area of the plastic part is given by the formula

$$A = 6x^2 + \frac{96}{x} \text{ cm}^2$$

4

(b) Find the minimum possible surface area that the plastic part could have.

7

[End of Question Paper]

Practice Paper 1

SQA Higher
Mathematics

Practice Paper 2 (Calculator)

Duration — 1 hour 45 minutes

Fill in these boxes and read what is printed below.

Full name of centre

Town

Candidate Forename(s)

Candidate Surname

Date of birth

Day	Month	Year		Scottish candidate number

Total marks available for this paper — 80

You may use a calculator.

Answer ALL the questions.

Write your answers in the spaces provided.

You must use **BLUE** or **BLACK** ink to write your answers.

To earn full marks you must show all of your working.

Give units with your answer where required.

You will find the formula list at the back of this book.

Total marks — 80

Attempt ALL questions

1. The point A (4, 7) lies on the circle
   with equation $x^2 + y^2 - 16x - 10y + 69 = 0$.

   Calculate the equation of the tangent
   to the circle at A.

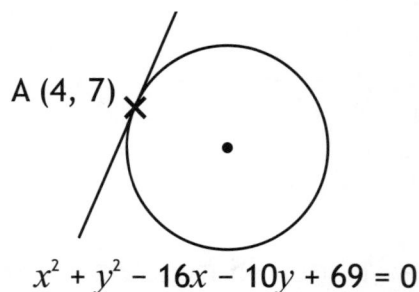

   A (4, 7)

   $x^2 + y^2 - 16x - 10y + 69 = 0$

   4

2. The function f(x) is defined as f$(x) = x^2 - 6x + 2$.

   Determine the range of values of x for which f$(x) > -6$.

   3

3. A sequence is generated by the recurrence relation

   $$u_{n+1} = (k-1)u_n + 3 \text{ where } k \text{ is a positive constant.}$$

   (a) Given that $u_3 = 2k + 2$, find an expression for u_4 in terms of k only.

   2

   (b) If $u_4 = 6.12$, calculate the value of k.

   1

   The sequence converges to a limit as n increases.

   (c) Calculate the limit of the sequence.

   2

4. The functions f(x), g(x) and h(x) are defined for all real numbers, where:

$$g(x) = af(bx) \qquad h(x) = f(x + c) + d$$

The graphs of $y = f(x)$, $y = g(x)$ and $y = h(x)$ are shown below.

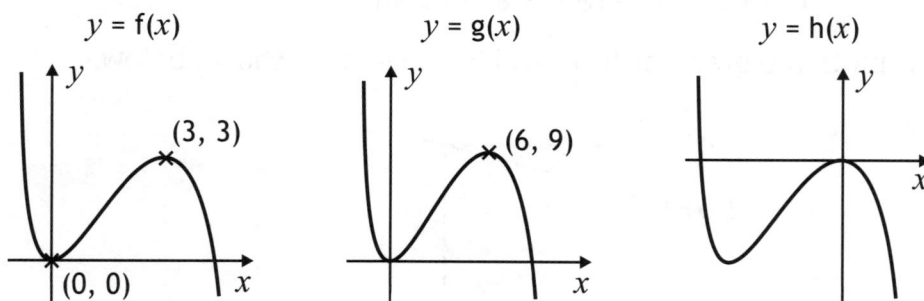

y = f(x)

(3, 3)

(0, 0)

y = g(x)

(6, 9)

y = h(x)

Write down the values of a, b, c and d.

4

5. The diagram shows the graph of $y = \dfrac{3}{4\sqrt{(5x + 1)^3}}$.

(a) Find the equation of the tangent to the curve at the point P, where the curve crosses the y-axis.

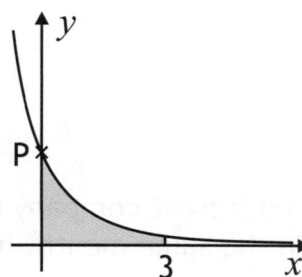

5

P

3

(b) Calculate the shaded area on the diagram, bounded by the coordinate axes, the curve, and the line $x = 3$.

5

6. An economist is analysing the value of a company named BankCorp.

The value, V, of the company, in thousands of pounds, t months after the company started, is modelled by the formula

$$V = V_0 e^{rt} \text{ where } r \text{ is a constant.}$$

The economist plots the graph of $\ln V$ against t, which is shown below.

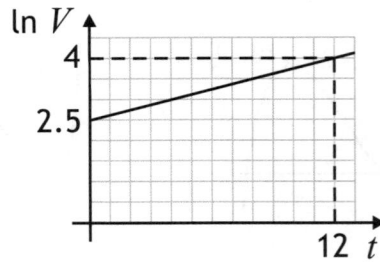

(a) Using the graph, work out the values of V_0 and r.

5

(b) A different company named Enterprise Inc., which started in the same month, has a value modelled by the formula $W = 20e^{0.08t}$

Calculate the value of t when BankCorp is first worth more than Enterprise Inc.

4

7. Given that $y = \cos^3 x$, find $\dfrac{dy}{dx}$.

2

8. Three circles, C_R, C_S and C_T have their centres at the points R (–13, 16), S (k, 10) and T (15, –5) respectively.

   (a) The points R, S and T are collinear.

      (i) Find the ratio in which the point S divides the line RT.   **1**

      (ii) Determine the value of k.   **1**

   It is given that:

   - C_R and C_T touch externally,

   - C_R and C_T both touch C_S internally,

   - $r_S = r_R + r_T$, where r_S, r_R and r_T are the radii of C_S, C_R and C_T respectively.

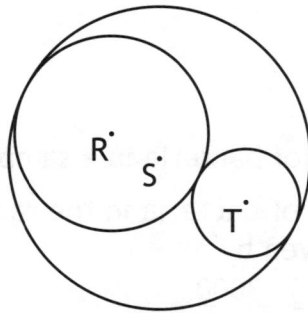

   (b) Determine the coordinates of the point where C_R and C_T meet.   **5**

9. Solve the equation $2\sin\left(x + \frac{\pi}{3}\right) = 3\cos\left(x - \frac{\pi}{2}\right)$
in the interval $0 \le x \le 2\pi$.

6

10. A scientist is studying the number of bacteria in a sample.

    The rate of change of the number of bacteria in the sample, n, t days after the study began, is given by
    $$\frac{dn}{dt} = \frac{3000}{(t+1)^4}$$
    At the beginning of the study, there were 500 bacteria in the sample.

    Find an expression for n in terms of t.

    4

11. The amplitude of a wave at time t milliseconds is given by

$$A(t) = 7 \cos t - 2 \sin t$$

(a) Write $A(t)$ in the form $p \sin(t + q)$, where $p > 0$ and $0 \le q \le 2\pi$.

4

(b) Find the value(s) of t when $A(t) = 1$ for $0 \le t \le 2\pi$.

4

12. The cuboid shown below has sides given by the vectors **r**, **s** and **t**, where **r** = (3**i** + 2**j** + 2**k**), **s** = (2**i** + m**j** + 4**k**) and **t** = (n**i** − 8**j** − 25**k**).

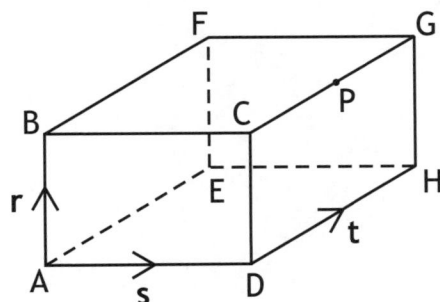

(a) Calculate the values of m and n.

4

(b) The point P is the midpoint of CG.
Express the vector $\overrightarrow{AP}$ in terms of **i**, **j** and **k**.

2

(c) Hence calculate the size, in degrees, of angle BAP.

13. The diagram below shows the graphs of $y = 3x^2 - 2x^3$ and $y = -2x$.

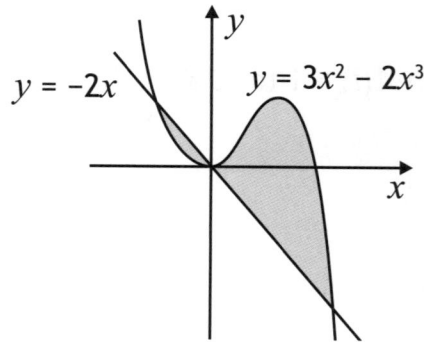

Calculate the total shaded area on the diagram.

7

[End of Question Paper]

Practice Paper 2

Section One — Algebraic Skills

Pages 4-5: Quadratic Equations

1 Calculate the discriminant:
$b^2 - 4ac = (-4)^2 - 4 \times 1 \times 2 = 16 - 8 = 8$ *[1 mark]*
The discriminant is positive,
so the equation has two distinct real roots. *[1 mark]*
[2 marks available in total — as above]

2 Expanding the brackets on the RHS gives the quadratic
$mx^2 + 4mx + 4m + p$.
Equating the coefficients of x^2 gives $m = 5$ *[1 mark]*
Equating the coefficients of x gives $n = 4m \Rightarrow n = 20$ *[1 mark]*
Equating the constant terms gives $14 = 4m + p \Rightarrow p = -6$ *[1 mark]*
[3 marks available in total — as above]

3 Rewrite $x^2 - 18x$ as one bracket squared
while adding a constant $d \Rightarrow (x - 9)^2 + d$ *[1 mark]*
So $x^2 - 18x + 16 = (x - 9)^2 + d$
$\Rightarrow x^2 - 18x + 16 = x^2 - 18x + 81 + d$
$\Rightarrow d = 16 - 81 = -65$
So $x^2 - 18x + 16 = (x - 9)^2 - 65$ *[1 mark]*
[2 marks available in total — as above]

4 If the quadratic $ax^2 + 2x + 2 = 0$ has one real root,
then the discriminant is 0, i.e. $b^2 - 4ac = 0$
$\Rightarrow (2)^2 - (4 \times a \times 2) = 0$ *[1 mark]*
$\Rightarrow 4 - (8a) = 0$
$\Rightarrow 8a = 4$
$\Rightarrow a = \frac{1}{2}$ *[1 mark]*
[2 marks available in total — as above]

5 If the quadratic $x^2 - 4x + (k - 1) = 0$ has no real roots,
then the discriminant is negative, i.e. $b^2 - 4ac < 0$
$\Rightarrow (-4)^2 - (4 \times 1 \times (k - 1)) < 0$ *[1 mark]*
$\Rightarrow 16 - (4k - 4) < 0$
$\Rightarrow 20 - 4k < 0$ *[1 mark]*
$\Rightarrow 20 < 4k$
$\Rightarrow k > 5$ *[1 mark]*
[3 marks available in total — as above]

6 a) Rewrite $x^2 - 7x$ as one bracket squared
while adding a constant d:
$\left(x - \frac{7}{2}\right)^2 + d$ *[1 mark]*
So $x^2 - 7x + 17 = \left(x - \frac{7}{2}\right)^2 + d$
$\Rightarrow x^2 - 7x + 17 = x^2 - 7x + \frac{49}{4} + d$
so $d = 17 - \frac{49}{4} = \frac{19}{4}$
So $x^2 - 7x + 17 = \left(x - \frac{7}{2}\right)^2 + \frac{19}{4}$ *[1 mark]*
[2 marks available in total — as above]

b) The maximum value of f(x) will be when the denominator
is as small as possible — so you want the minimum value of
$x^2 - 7x + 17$. Using the completed square above, the minimum
value is $\frac{19}{4}$ as the squared part is greater than or equal to 0.
So the maximum value of f(x) is $\frac{1}{\left(\frac{19}{4}\right)} = \frac{4}{19}$ or 0.211 (3 s.f.)

*[2 marks available — 1 mark for justification of finding
the minimum of the quadratic, 1 mark for correct answer]*

7 If the quadratic $3x^2 + kx + 2k = 0$ has one real root,
then the discriminant is 0, i.e. $b^2 - 4ac = 0$
$\Rightarrow k^2 - (4 \times 3 \times 2k) = 0$ *[1 mark]*
$\Rightarrow k^2 - 24k = 0$ *[1 mark]*
$\Rightarrow k(k - 24) = 0$
$\Rightarrow k = 0$ or $k = 24$ *[1 mark for both]*
[3 marks available in total — as above]

Page 6: Quadratic Inequalities

1 $3x^2 - 5x - 2 \leq 0 \Rightarrow (3x + 1)(x - 2) \leq 0$
Sketch a graph to see where the quadratic is less than or equal to 0
— it'll be a u-shaped curve that crosses the x-axis at $x = -\frac{1}{3}$ and $x = 2$.

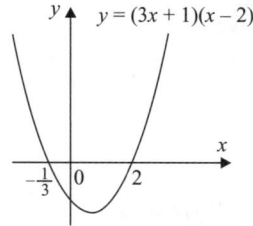

From the graph, the quadratic is negative when $-\frac{1}{3} \leq x \leq 2$.
*[2 marks available — 1 mark for finding the roots,
1 mark for the correct range with justification]*

2 The area of the office will be $(x - 9)(x - 6)$ m^2, so use this to form an
inequality for the necessary floor space:
$(x - 9)(x - 6) \geq 28 \Rightarrow x^2 - 15x + 54 \geq 28$
$\Rightarrow x^2 - 15x + 26 \geq 0$
Find the x-intercepts of the graph:
$x^2 - 15x + 26 = 0 \Rightarrow (x - 2)(x - 13) = 0$
$\Rightarrow x = 2$ or $x = 13$

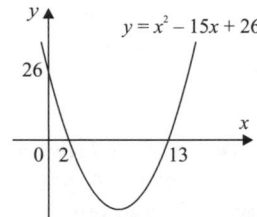

From the graph, $x^2 - 15x + 26 \geq 0 \Rightarrow x \leq 2$ or $x \geq 13$
But $x \leq 2$ would mean that the sides of
the office would have negative lengths,
so the only possible values of x are $x \geq 13$ metres.
*[3 marks available — 1 mark for forming a quadratic,
1 mark for finding the roots, 1 mark for the correct
range with justification]*

3 a) The volume of cuboid A is $5x(x + 1) = 5x^2 + 5x$
The volume of cuboid B is $6x^2$ *[1 mark for both]*
$5x^2 + 5x > 6x^2$ so $x^2 - 5x < 0$ *[1 mark]*
[2 marks available in total — as above]

b) $x^2 - 5x < 0 \Rightarrow x(x - 5) < 0$
Sketch a graph to see where the quadratic is greater than 0 — it'll
be a u-shaped curve that crosses the x-axis at $x = 0$ and $x = 5$.

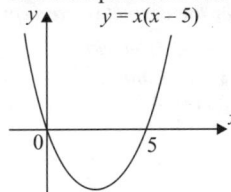

From the graph, the quadratic is negative when $0 < x < 5$.
Since x is an integer greater than 1,
the possible values of x are $x = 2$, 3 or 4.
*[3 marks available — 1 mark for finding the roots,
1 mark for the correct range with justification,
1 mark for listing the correct values only]*

76

Pages 7-9: Cubics and Quartics

1 a) If $(x-3)$ is a factor then f(3) = 0.
f(3) = $(3)^3 - 6(3)^2 - (3) + 30$ *[1 mark]*
= 27 − 54 − 3 + 30 = 0
So $(x-3)$ is a factor of f(x). *[1 mark]*
[2 marks available in total — as above]

b) Use synthetic division to divide $x^3 - 6x^2 - x + 30$ by $(x-3)$:

3	1	−6	−1	30
		3	−9	−30
	1	−3	−10	0

So $x^3 - 6x^2 - x + 30 = (x-3)(x^2 - 3x - 10)$ *[1 mark]*
Factorising the quadratic expression gives:
f(x) = $(x-3)(x-5)(x+2)$ *[1 mark]*
[2 marks available in total — as above]

2 a) The cubic has already been factorised, so the curve
crosses the x-axis at (1, 0), (−2, 0) and (3, 0).
When x = 0, y = (−1)(2)(3) = −6, so the y-intercept is (0, −6).
The coefficient of the x^3 term is negative (−1), so the cubic goes
from top left to bottom right. The graph looks like this:

*[3 marks available — 1 mark for the correct shape, 1 mark for
the correct x-intercepts, 1 mark for the correct y-intercept]*

b) The quartic has already been factorised — there are two
double roots, one at (2, 0) and the other at (−3, 0). When x = 0,
y = $(-2)^2(3^2)$ = 36, so the y-intercept is (0, 36). The coefficient of
the x^4 term is positive, and as the graph only touches the
x-axis but doesn't cross it, it is always above the x-axis.
The graph looks like this:

*[3 marks available — 1 mark for the correct shape, 1 mark for
the correct x-intercepts, 1 mark for the correct y-intercept]*

3 The repeated root at x = −2 means that t = −2 *[1 mark]*
and the single root at x = 2 means that s = 2 *[1 mark]*
The graph passes through (1, −4.5) so f(1) = −4.5
$\Rightarrow$ f(1) = $r(1)(1-2)(1+2)^2$ = −4.5
$\Rightarrow$ $r(1)(-1)(3)^2$ = −4.5
$\Rightarrow$ −9r = −4.5 $\Rightarrow$ r = 0.5 *[1 mark]*
[3 marks available in total — as above]

4 By the remainder theorem, if the remainder when f(x)
is divided by $(x-a)$ is b, then f(a) = b. *[1 mark]*
The remainder when f(x) is divided by $(x-3)$ is 7 $\Rightarrow$ f(3) = 7
$\Rightarrow$ $(3)^3 + a(3)^2 + b(3) + 7$ = 7
$\Rightarrow$ 27 + 9a + 3b + 7 = 7
$\Rightarrow$ 9a + 3b + 27 = 0 *[1 mark]*
The remainder when f(x) is divided by $(x+2)$ is −3 $\Rightarrow$ f(−2) = −3
$\Rightarrow$ $(-2)^3 + a(-2)^2 + b(-2) + 7$ = −3
$\Rightarrow$ −8 + 4a − 2b + 7 = −3
$\Rightarrow$ 2b = 4a + 2 $\Rightarrow$ b = 2a + 1 *[1 mark]*
E.g. substituting b = 2a + 1 into 9a + 3b + 27 = 0:
9a + 3(2a + 1) + 27 = 0
$\Rightarrow$ 9a + 6a + 3 + 27 = 0
$\Rightarrow$ 15a + 30 = 0 $\Rightarrow$ a = −2 *[1 mark]*
Then b = 2a + 1 $\Rightarrow$ b = 2(−2) + 1 = −3 *[1 mark]*
[5 marks available in total — as above]

5 a) If $(2x + 1)$ is a factor then $f\left(-\frac{1}{2}\right) = 0$.
$f\left(-\frac{1}{2}\right) = 2\left(-\frac{1}{2}\right)^3 - \left(-\frac{1}{2}\right)^2 - 5\left(-\frac{1}{2}\right) - 2$ *[1 mark]*
$= -\frac{2}{8} - \frac{1}{4} + \frac{5}{2} - 2 = 0,$
so $(2x + 1)$ is a factor of f(x). *[1 mark]*
[2 marks available in total — as above]

b) Use synthetic division to divide $2x^3 - x^2 - 5x - 2$ by $(2x + 1)$:

$-\frac{1}{2}$	2	−1	−5	−2
		−1	1	2
	2	−2	−4	0

So $2x^3 - x^2 - 5x - 2 = \left(x + \frac{1}{2}\right)(2x^2 - 2x - 4)$ *[1 mark]*
$= \left(x + \frac{1}{2}\right) \times 2 \times (x^2 - x - 2)$
$= (2x + 1)(x^2 - x - 2)$
Factorising the quadratic expression gives:
f(x) = $(2x + 1)(x - 2)(x + 1)$ *[1 mark]*
[2 marks available in total — as above]

6 If $(x - 1)$ is a factor of f(x) then f(1) = 0. *[1 mark]*
f(1) = $1^3 - 2(1)^2 - a(1) + 12$, so 0 = 11 − a $\Rightarrow$ a = 11 *[1 mark]*
So f(x) = $x^3 - 2x^2 - 11x + 12$.
To solve f(x) = 0, first factorise $x^3 - 2x^2 - 11x + 12$. $(x - 1)$ is a factor,
so use synthetic division to divide $x^3 - 2x^2 - 11x + 12$ by $(x - 1)$:

1	1	−2	−11	12
		1	−1	−12
	1	−1	−12	0

So $x^3 - 2x^2 - 11x + 12 = (x - 1)(x^2 - x - 12)$ *[1 mark]*
Then factorise the quadratic:
$= (x - 1)(x - 4)(x + 3)$ *[1 mark]*
Finally, solve f(x) = 0:
$x^3 - 2x^2 - 11x + 12 = 0 \Rightarrow (x - 1)(x - 4)(x + 3) = 0,$
so x = 1, x = 4 or x = −3 *[1 mark]*
[5 marks available in total — as above]

7 The curve intersects the x-axis at −5, −3, −2 and z,
so the function factorises as f(x) = $(x + 5)(x + 3)(x + 2)(x - z)$. *[1 mark]*
f(x) passes through (0, −120), so:
f(0) = $5 \times 3 \times 2 \times -z = -120$ *[1 mark]*
$\Rightarrow$ −30z = −120 $\Rightarrow$ z = 4 *[1 mark]*
[3 marks available in total — as above]

8 a) Using the remainder theorem: f(−1) = −24 *[1 mark]*
$\Rightarrow$ $(-1)^4 - 2(-1)^3 - 9(-1)^2 + s(-1) = -24$
$\Rightarrow$ 1 + 2 − 9 − s = −24 $\Rightarrow$ s = 18 *[1 mark]*
[2 marks available in total — as above]

b) Use synthetic division to divide $x^4 - 2x^3 - 9x^2 + 18x$ by $(x + 3)$:

−3	1	−2	−9	18	0
		−3	15	−18	0
	1	−5	6	0	0

So $x^4 - 2x^3 - 9x^2 + 18x = (x + 3)(x^3 - 5x^2 + 6x)$ *[1 mark]*
$= (x + 3) \times x \times (x^2 - 5x + 6)$ *[1 mark]*
$= x(x + 3)(x - 3)(x - 2)$ *[1 mark]*
[3 marks available in total — as above]

Answers

Pages 10-11: Graph Transformations

1 a) $4 - 4^x = -4^x + 4$ so the graph has been reflected in the x-axis and then translated up by 4 units.

When $x = 0$, $4 - 4^0 = 3$, so $y = 4 - 4^x$ intersects the y-axis at $(0, 3)$.

When $y = 0$, $4 - 4^x = 0 \Rightarrow 4 = 4^x \Rightarrow x = 1$, so $y = 4 - 4^x$ intersects the x-axis at $(1, 0)$

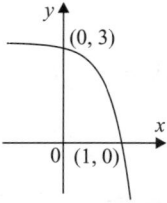

[3 marks available — 1 mark for the correct shape, 1 mark for the correct y-intercept, 1 mark for the correct x-intercept]

2 a) Find where the graph of $y = f(x)$ touches the x-axis:

When $f(x) = 0$, $(x - 1)^2(x + 2)^2 = 0 \Rightarrow x = 1$ or $x = -2$ *[1 mark]*

The graph of $f(x - 3)$ is the graph of $f(x)$ translated 3 units to the right, so the curve will touch the x-axis at

$x = 1 + 3 = 4$ and $x = -2 + 3 = 1$. *[1 mark]*

[2 marks available in total — as above]

b) The graph of $y = f(x)$ touches, but never goes below the x-axis. The graph of $f(x) + c$ is the graph of $f(x)$ translated up by c units, so $g(x)$ will have no real roots if $c > 0$. *[1 mark]*

3 a) The graph of $y = 3f(x + 2)$ is a translation of the graph of $y = f(x)$ by 2 units to the left, followed by a stretch in the y-direction by a factor of 3.

So to find the new coordinates, subtract 2 from the x-coordinate and multiply the y-coordinate by 3:

$(-1, -2)$ will move to: $(-1 - 2, -2 \times 3) = (-3, -6)$ *[1 mark]*

$(3, 2)$ will move to: $(3 - 2, 2 \times 3) = (1, 6)$ *[1 mark]*

[2 marks available in total — as above]

b) The graph of $y = g(x)$ is a reflection of the graph of $y = 3f(x + 2)$ in the x-axis, so to find the new coordinates, multiply the y-coordinates by -1:

$(-3, -6)$ will move to: $(-3, -6 \times -1) = (-3, 6)$

$(1, 6)$ will move to: $(1, 6 \times -1) = (1, -6)$

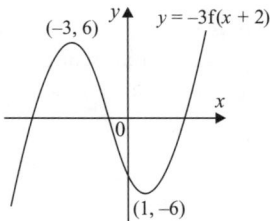

[2 marks available — 1 mark for the correct shape and orientation, 1 mark for the new turning points correctly labelled]

4 a) The graph of $y = f(x)$ has been translated $5 - 2 = 3$ units to the left and $13 - 3 = 10$ units down.

So $a = 3$ *[1 mark]* and $b = -10$ *[1 mark]*

[2 marks available in total — as above]

b) From part a) you know that the image of the point P on the graph of $y = g(x)$ has the coordinates $(X - 3, Y - 10)$ *[1 mark]*

The graph of $y = 2g(x + 1)$ is the graph of $y = g(x)$ shifted 1 unit to the left and then stretched vertically by a factor of 2, so subtract 1 from the x-coordinate and double the y-coordinate of the image of the point P to get the coordinates $(X - 4, 2Y - 20)$ *[1 mark]*

[2 marks available in total — as above]

Pages 12-15: Exponentials and Logs

1 Use the laws of logs to simplify:

$\dfrac{\ln 54 - \ln 6}{\ln 3} = \dfrac{\ln(54 \div 6)}{\ln 3}$ *[1 mark]*

$= \dfrac{\ln 9}{\ln 3} = \dfrac{\ln 3^2}{\ln 3}$ *[1 mark]*

$= \dfrac{2\ln 3}{\ln 3} = 2$ *[1 mark]*

[3 marks available in total — as above]

2 Use the laws of logs to simplify:

$\log_3 243 - \dfrac{1}{2}\log_3 9 = \log_3 243 - \log_3 9^{\frac{1}{2}}$ *[1 mark]*

$= \log_3 243 - \log_3 3$

$= \log_3 \dfrac{243}{3}$ *[1 mark]*

$= \log_3 81 = 4$ *[1 mark]*

[3 marks available in total — as above]

3 Rewrite the expression using the laws of logs:

$\log_a 4 + 3\log_a 2 = \log_a 4 + \log_a 2^3$ *[1 mark]*

$= \log_a (4 \times 2^3)$ *[1 mark]*

$= \log_a 32$

Therefore $\log_a x = \log_a 32$, so $x = 32$ *[1 mark]*

[3 marks available in total — as above]

4 a) $f(x) = c^x$, so $64 = c^3$ *[1 mark]* $\Rightarrow c = \sqrt[3]{64} = 4$ *[1 mark]*

[2 marks available in total — as above]

b)

[2 marks available — 1 mark for the correct shape, 1 mark for both correct coordinates]

5 Rewrite the expression using the laws of logs:

$\log_a 54 + \log_a 4 = \log_a (54 \times 4)$ *[1 mark]*

$= \log_a 216$

$\log_a 216 = 3 \Rightarrow a = \sqrt[3]{216} = 6$, so $a = 6$ *[1 mark]*

[2 marks available in total — as above]

6 Rewrite all the terms as powers of p and use the laws of logs to simplify:

$\log_p (p^4) + \log_p \left(p^{\frac{1}{2}}\right) - \log_p \left(p^{-\frac{1}{2}}\right)$ *[1 mark]*

$= 4\log_p p + \dfrac{1}{2}\log_p p - \left(-\dfrac{1}{2}\right)\log_p p$ *[1 mark]*

$= 4 + \dfrac{1}{2} - \left(-\dfrac{1}{2}\right) = 4 + 1 = 5$ (as $\log_p p = 1$) *[1 mark]*

[3 marks available in total — as above]

7 a) $\log_4 p - \log_4 q = \dfrac{1}{2}$, so using the log laws:

$\log_4 \left(\dfrac{p}{q}\right) = \dfrac{1}{2}$ *[1 mark]*

$\dfrac{p}{q} = 4^{\frac{1}{2}}$ *[1 mark]* $= \sqrt{4} = 2 \Rightarrow p = 2q$ *[1 mark]*

[3 marks available in total — as above]

b) Since $p = 2q$ (from a)), the equation can be written:

$\log_2 (2q) + \log_2 q = 7$ *[1 mark]*

$\Rightarrow \log_2 (2q \times q) = 7 \Rightarrow \log_2 (2q^2) = 7$ *[1 mark]*

$\Rightarrow 2q^2 = 2^7 = 128$ *[1 mark]*

$\Rightarrow q^2 = 64 \Rightarrow q = 8$ (since p and q are positive) *[1 mark]*

Then $p = 2q \Rightarrow p = 16$ *[1 mark]*

[5 marks available in total — as above]

8   For $3\ln x - \ln 3x = 0$, use the laws of logs to simplify to:
$\ln x^3 = \ln 3x$ *[1 mark]*
$\Rightarrow x^3 = 3x$ *[1 mark]*
$\Rightarrow x(x^2 - 3) = 0$
$x = 0, x = \pm\sqrt{3}$
$x > 0$ so $x = \sqrt{3}$ is the only solution. *[1 mark]*
[3 marks available in total — as above]

9   a) $V = pq^t \Rightarrow \log V = \log pq^t$ *[1 mark]*
$\Rightarrow \log V = \log p + t\log q$ *[1 mark]*
This is the equation of a straight line, where
$\log p$ is the intercept and $\log q$ is the gradient. *[1 mark]*
So $\log p = 4 \Rightarrow p = 10^4$ *[1 mark]*
and $\log q = -0.025 \Rightarrow q = 10^{-0.025}$ *[1 mark]*
[5 marks available in total — as above]

   b)   Using $V = pq^t$ when $t = 20$:
$V = (10^4)(10^{-0.025})^{20}$ *[1 mark]*
$= (10^4)(10^{-0.5}) = £3162$ (nearest £) *[1 mark]*
[2 marks available in total — as above]

10 a) $24 = 20e^{b(1)} \Rightarrow e^b = 1.2 \Rightarrow \ln e^b = \ln 1.2$ *[1 mark]*
$\Rightarrow b = \ln 1.2 = 0.1823... = 0.182$ (3 s.f.) *[1 mark]*
[2 marks available in total — as above]

   b)   Substitute H for 500 and solve for t:
$500 = 20e^{(0.1823... \times t)}$ *[1 mark]*
$\Rightarrow 25 = e^{(0.1823... \times t)} \Rightarrow \ln 25 = \ln e^{(0.1823... \times t)}$
$\Rightarrow \ln 25 = 0.1823... \times t$ *[1 mark]*
$\Rightarrow t = \dfrac{\ln 25}{0.1823...} = 17.654... = 17.7$ hours (3 s.f.) *[1 mark]*
[3 marks available in total — as above]

   c)   At $t = 7$, $H = 20e^{(\ln 1.2)7} = 71.66...$ *[1 mark]*
$\dfrac{71.66...}{2000} \times 100\% = 3.583...\% = 3.58\%$ (3 s.f.) *[1 mark]*
[2 marks available in total — as above]

11 $p = at^b \Rightarrow \log p = \log at^b$ *[1 mark]*
$\log p = \log a + b\log t$ *[1 mark]*
This is the equation of a straight line, where
$\log a$ is the intercept and b is the gradient. *[1 mark]*
So $\log a = 0.3 \Rightarrow a = 10^{0.3} = 1.995... = 2.0$ (1 d.p.) *[1 mark]*
and $b = \dfrac{1.3 - 0.3}{0.6} = \dfrac{1}{0.6} = 1.666... = 1.7$ (1 d.p.) *[1 mark]*
[5 marks available in total — as above]

12 a)   At the start of 2010, $t = 0 \Rightarrow P = 5700$
20% of 5700 = 1140 *[1 mark]*
When $P < 1140$, $1140 > 5700e^{-0.15t}$ *[1 mark]*
$\Rightarrow e^{-0.15t} < \dfrac{1140}{5700} = 0.2$ *[1 mark]*
Take ln of both sides: $-0.15t < \ln 0.2$ *[1 mark]*
$\Rightarrow t > -\dfrac{\ln 0.2}{0.15} = 10.729...$ years.
So the population will drop below 20% when $t > 10.7$ (3 s.f.)
i.e. in the year 2020. *[1 mark]*
[5 marks available in total — as above]

   b)   You need to find t such that:
$2100 - 1500e^{-0.15t} > 5700e^{-0.15t}$ *[1 mark]*
$\Rightarrow 2100 > 7200e^{-0.15t} \Rightarrow \dfrac{7}{24} > e^{-0.15t}$ *[1 mark]*
$\Rightarrow \ln \dfrac{7}{24} > \ln e^{-0.15t} \Rightarrow \ln \dfrac{7}{24} > -0.15t$ *[1 mark]*
$\ln \dfrac{7}{24} \div -0.15 < t$ *[1 mark]*
$t > 8.21429...$
So Q first exceeds P when $t > 8.21$ (3 s.f.),
i.e. in the year 2018. *[1 mark]*
[4 marks available in total — as above]

Pages 16-17: Composite and Inverse Functions

1   $f(f(x)) = f(4x^2)$ *[1 mark]*
so $f(4x^2) = 4(4x^2)^2 = 4 \times 16x^4 = 64x^4$ *[1 mark]*
So $64x^4 = 4 \Rightarrow x^4 = \dfrac{4}{64} = \dfrac{1}{16} \Rightarrow x = \dfrac{1}{2}$ (since $x \geq 0$) *[1 mark]*
[3 marks available in total — as above]

2   a) $g(f(x)) = g(2x^2 + 3)$
$g(2x^2 + 3) = \sqrt{2(2x^2 + 3) - 6}$ *[1 mark]*
$= \sqrt{4x^2 + 6 - 6} = \sqrt{4x^2} = 2x$ (since $x > 3$) *[1 mark]*
[2 marks available in total — as above]

   b) $f(g(x)) = f(\sqrt{2(x) - 6})$
$f(\sqrt{2(x) - 6}) = 2(\sqrt{2x - 6})^2 + 3$
$= 2(2x - 6) + 3 = 4x - 9$ *[1 mark]*
So $f(g(4)) = 4(4) - 9 = 7$ *[1 mark]*
[2 marks available in total — as above]

3   a) $g(f(x)) = g(x^2 + 2x - 8)$
$= \dfrac{1}{2(x^2 + 2x - 8)} = \dfrac{1}{2x^2 + 4x - 16}$ *[1 mark]*

   b)   The denominator of $g(f(x))$ cannot equal zero, so set the
denominator equal to zero and solve for x:
$2x^2 + 4x - 16 = 0 \Rightarrow (2x - 4)(x + 4) = 0$ *[1 mark]*
$\Rightarrow x = 2$ or -4
So $x = 2$ or -4 cannot be in the domain of $g(f(x))$. *[1 mark]*
[2 marks available in total — as above]

4   a) $g(f(x)) = x \Rightarrow g(x)$ is the inverse of $f(x)$.
Write $y = f(x)$ and rearrange to make x the subject:
$y = \dfrac{3}{2x + 5}$ *[1 mark]*
$\Rightarrow 2x + 5 = \dfrac{3}{y} \Rightarrow 2x = \dfrac{3}{y} - 5 \Rightarrow x = \dfrac{3}{2y} - \dfrac{5}{2}$ *[1 mark]*
Replace x with $g(x)$ and y with x:
$g(x) = \dfrac{3}{2x} - \dfrac{5}{2}$ *[1 mark]*
[3 marks available in total — as above]

   b)   f and g are inverse functions, so $f(2) = \dfrac{1}{3} \Rightarrow g\left(\dfrac{1}{3}\right) = 2$ *[1 mark]*

   c) $g(3) = \dfrac{3}{2(3)} - \dfrac{5}{2}$ *[1 mark]*
$= \dfrac{1}{2} - \dfrac{5}{2} = -2$ *[1 mark]*
[2 marks available in total — as above]

5   a)   First write $y = g(x)$ and rearrange to make x the subject:
$y = \sqrt{3x + 1}$ *[1 mark]* $\Rightarrow y^2 = 3x + 1 \Rightarrow y^2 - 1 = 3x$
$\Rightarrow \dfrac{y^2 - 1}{3} = x$ *[1 mark]*
Then replace x with $g^{-1}(x)$ and y with x:
$g^{-1}(x) = \dfrac{x^2 - 1}{3}$ *[1 mark]*
[3 marks available in total — as above]

   b) $f(0) = 2(0) + 1 = 1$ *[1 mark]*
$\Rightarrow g(f(0)) = \sqrt{3(1) + 1} = \sqrt{4} = 2$ *[1 mark]*
[2 marks available in total — as above]

   c) $g(f(x)) = g(2\sin x + 1)$ *[1 mark]*
$= \sqrt{3(2\sin x + 1) + 1} = \sqrt{6\sin x + 4}$ *[1 mark]*
[2 marks available in total — as above]

Answers

Pages 18-19: Recurrence Relations

1 a) Each term is 4 more than the previous term.
The first term is 12, so the recurrence relation
is $u_{n+1} = u_n + 4$ *[1 mark]*, $u_0 = 12$. *[1 mark]*
[2 marks available in total — as above]

b) A recurrence relation $u_{n+1} = au_n + b$ will only
give a convergent sequence if $-1 < a < 1$. In this case,
$a = 1$, so the sequence will not converge. *[1 mark]*

2 Substitute $a = \frac{1}{3}$ and $b = 2$ into the limit formula:

$$L = \frac{b}{1-a} = \frac{2}{1-\frac{1}{3}} = \frac{2}{\frac{2}{3}} = 3$$

You could also write $L = \frac{1}{3}L + 2$ and then solve for L.

[2 marks available — 1 mark for a correct method,
1 mark for the correct answer]

3 a) Substitute $u_0 = 4$ into the formula $u_{n+1} = xu_n - 12$:
$u_1 = 4x - 12$ *[1 mark]*
Substitute $u_1 = 4x - 12$ into the formula:
$u_2 = x(4x - 12) - 12 = 4x^2 - 12x - 12$ *[1 mark]*
[2 marks available in total — as above]

b) If $u_0 = u_2$, $4x^2 - 12x - 12 = 4$ *[1 mark]*
$\Rightarrow 4x^2 - 12x - 16 = 0$ *[1 mark]*
$\Rightarrow x^2 - 3x - 4 = 0$
$\Rightarrow (x + 1)(x - 4) \Rightarrow x = -1$ and $x = 4$ *[1 mark]*
[3 marks available in total — as above]

4 $u_2 = au_1 + b \Rightarrow 7 = 6a + b$
$u_3 = au_2 + b \Rightarrow 8.5 = 7a + b$ *[1 mark for both]*
$\Rightarrow 8.5 - 7 = 7a - 6a \Rightarrow a = 1.5$ *[1 mark]*
and $b = 7 - 6(1.5) = -2$ *[1 mark]*
[3 marks available in total — as above]

5 a) $u_0 = 30$, $u_1 = 525$ and $u_2 = 921$, so:
$u_1 = au_0 + b \Rightarrow 525 = 30a + b$ *[1 mark]*
$u_2 = au_1 + b \Rightarrow 921 = 525a + b$ *[1 mark]*
Solve the equations simultaneously for a:
$525a + b = 921$
$\underline{30a + b = 525 -}$
$495a = 396$
$\Rightarrow a = \frac{4}{5}$ *[1 mark]*
Put a back into an equation to find b:
$30 \times \frac{4}{5} + b = 525 \Rightarrow b = 501$ *[1 mark]*
[4 marks available in total — as above]

b) Substitute $a = \frac{4}{5}$ and $b = 501$ into the limit formula:
$$L = \frac{b}{1-a} = \frac{501}{1-\frac{4}{5}} = \frac{501}{\frac{1}{5}} = 501 \times 5 = 2505$$
[2 marks available — 1 mark for a correct method,
1 mark for the correct answer]

6 a) 2011 is one year later than 2010, so:
Town A: $u_1 = 0.45(37\%) + 27\% = 43.65\%$ *[1 mark]*
Town B: $v_1 = 0.71(40\%) + 15\% = 43.4\%$ *[1 mark]*
43.65% > 43.4%, so Town A had the higher recycling rate.
[1 mark]
[3 marks available in total — as above]

b) Town A:
$$L = \frac{b}{1-a} = \frac{27}{1-0.45}$$ *[1 mark]*
$$= \frac{27}{0.55} = 49.09...\%$$
So Town A will not meet the target. *[1 mark]*
Town B:
$$L = \frac{b}{1-a} = \frac{15}{1-0.71}$$ *[1 mark]*
$$= \frac{15}{0.29} = 51.72...\%$$
So Town B will meet the target. *[1 mark]*
[4 marks available in total — as above]

Section Two — Trigonometric Skills

Pages 20–21 — Trig Graphs and Transformations

1 The function $x(t)$ takes values between 2 and –4, so its amplitude is 3,
and $\sin t$ has an amplitude of 1. This means $x(t)$ has been stretched
vertically by a factor of 3, so $a = 3$. *[1 mark]*
$x(t)$ completes 1 'cycle' in π radians, and $\sin t$ completes 1 'cycle'
in 2π radians. This means $x(t)$ has been squashed horizontally by a
factor of 2, so $b = 2$. *[1 mark]*
$x(t)$ intersects the y-axis at –1, and $\sin t$ intersects the y-axis at 0.
This means $x(t)$ has been translated down by 1, so $c = -1$. *[1 mark]*
[3 marks available in total — as above]

2 The period of $a\cos bx$ is given by $T = \frac{2\pi}{b}$,
so the period of $8\cos 6x$ is $T = \frac{2\pi}{6} = \frac{\pi}{3}$.
[2 marks available — 1 mark for a correct method,
1 mark for the correct answer]

3 a) The maximum of the function f(x) occurs when $\cos x = -1$,
i.e. when $x = \pi$. *[1 mark]*

b) $g(x) = 6 - 4\cos\left(x - \frac{\pi}{6}\right)$.
This is at its minimum when $\cos\left(x - \frac{\pi}{6}\right) = 1$. *[1 mark]*
$6 - 4(1) = 2$, so the minimum of g(x) is 2. *[1 mark]*
[2 marks available in total — as above]

4 a)

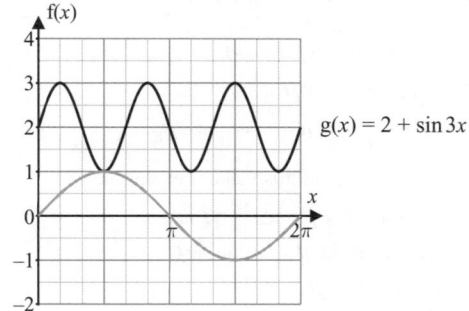

[2 marks available — 1 mark for correctly translating
the graph up by 2, 1 mark for squashing the graph
horizontally by factor of 3]

b) Period $= 2\pi \div 3 = \frac{2\pi}{3}$ *[1 mark]*
You could also read the period straight off the graph.

5 a) $\cos t$ has a maximum of 1 and a minimum of –1.
You want A + B $\cos t$ to have a maximum of 17
and a minimum of 7, so form two equations:
① A + B(1) = 17 (at maximum)
② A + B(–1) = 7 (at minimum)
①+②: 2A = 24 $\Rightarrow$ A = 12
Sub back into ①: 12 + B = 17 $\Rightarrow$ B = 5
This is a transformation of the graph of $\cos t$ — a vertical stretch
with scale factor 5 followed by a translation up by 12.
[3 marks available — 1 mark for a correct method,
1 for the correct value of A, 1 mark for the correct value of B]

b) When $t = 0$, f(t) is at its maximum of 17.
This will occur when $\cos(Ct + D) = 1$, i.e. when $Ct + D = 0$.
So $C(0) + D = 0 \Rightarrow D = 0$ *[1 mark]*
When $t = 6$, f(t) is at its minimum of 7.
This will occur when $\cos(Ct + D) = -1$ i.e. when $Ct + D = \pi$.
So $C(6) + D = \pi \Rightarrow 6C + 0 = \pi \Rightarrow C = \frac{\pi}{6}$ *[1 mark]*
Because cos is periodic, you might have found $D = 2\pi$.
Also, because cos is symmetrical, $C = -\frac{\pi}{6}$ will give you
the same answer — either one is fine.
[2 marks available in total — as above]

Answers

Pages 22-23 — Solving Trig Equations

1 Consider the following triangle:

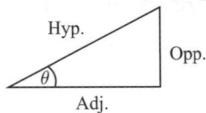

You are given that $\cos\theta = \dfrac{5}{6} \Rightarrow \dfrac{\text{Adj.}}{\text{Hyp.}} = \dfrac{5}{6}$

So if Adj. = 5 then Hyp. = 6. Using Pythagoras' Theorem:

Opp. $= \sqrt{6^2 - 5^2} = \sqrt{36 - 25} = \sqrt{11}$

So $\sin\theta = \dfrac{\text{Opp.}}{\text{Hyp.}} = \dfrac{\sqrt{11}}{6}$ and $\tan\theta = \dfrac{\text{Opp.}}{\text{Adj.}} = \dfrac{\sqrt{11}}{5}$

[3 marks available — 1 mark for a correct method, 1 mark for correct value of sin, 1 mark for the correct value of tan]
You could also use the fact that $\sin^2 x + \cos^2 x = 1$ to find $\sin x$, and use the fact that $\sin x \div \cos x = \tan x$ to find $\tan x$.

2 a) Use the trig identity $\tan\theta \equiv \dfrac{\sin\theta}{\cos\theta}$:

$\tan^2\theta + \dfrac{\tan\theta}{\cos\theta} = 1 \Rightarrow \dfrac{\sin^2\theta}{\cos^2\theta} + \dfrac{\sin\theta}{\cos^2\theta} = 1$ *[1 mark]*

$\Rightarrow \dfrac{\sin^2\theta + \sin\theta}{\cos^2\theta} = 1 \Rightarrow \sin^2\theta + \sin\theta = \cos^2\theta$ *[1 mark]*

Now use the identity $\cos^2\theta \equiv 1 - \sin^2\theta$ to give:

$\sin^2\theta + \sin\theta = 1 - \sin^2\theta$ *[1 mark]*

$\Rightarrow 2\sin^2\theta + \sin\theta - 1 = 0$ *[1 mark]*

[4 marks available in total — as above]

b) Factorising the quadratic from a) gives:
$(2\sin\theta - 1)(\sin\theta + 1) = 0$ *[1 mark]*

$\Rightarrow \sin\theta = \dfrac{1}{2}$ or $\sin\theta = -1$ *[1 mark]*

$\sin\theta = \dfrac{1}{2} \Rightarrow \theta = \dfrac{\pi}{6}$ and $\theta = \left(\pi - \dfrac{\pi}{6}\right) = \dfrac{5\pi}{6}$

$\sin\theta = -1 \Rightarrow \theta = \dfrac{3\pi}{2}$, but this value is excluded.

So the solutions are $\theta = \dfrac{\pi}{6}$ and $\dfrac{5\pi}{6}$ *[1 mark]*

[3 marks available in total — as above]

3 Consider the following triangle:

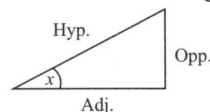

You are given that $\sin x = \dfrac{8}{9} \Rightarrow \dfrac{\text{Opp.}}{\text{Hyp.}} = \dfrac{8}{9}$

So if Opp. = 8 then Hyp. = 9. Using Pythagoras' Theorem:

Adj. $= \sqrt{9^2 - 8^2} = \sqrt{81 - 64} = \sqrt{17}$ *[1 mark]*

So $\tan x = \dfrac{\text{Opp.}}{\text{Adj.}} = \dfrac{8}{\sqrt{17}}$ *[1 mark]* $\Rightarrow \tan^2 x = \dfrac{64}{17}$ *[1 mark]*

[3 marks available in total — as above]

4 Rearrange the equation to get in terms of one trig function:
$4\sin 2x - \cos 2x = 0 \Rightarrow 4\sin 2x = \cos 2x$

$\Rightarrow \dfrac{\sin 2x}{\cos 2x} = \dfrac{1}{4} \Rightarrow \tan 2x = \dfrac{1}{4}$ *[1 mark]*

So solve $\tan 2x = \dfrac{1}{4}$ in the range $0 \le 2x \le 4\pi$:

$2x = \tan^{-1}\dfrac{1}{4} = 0.244..., 3.386..., 6.528..., 9.669...$ *[1 mark]*

So $x = 0.122, 1.693, 3.264, 4.835$ (all to 3 d.p.)

[1 mark for all 4 correct solutions only]
[3 marks available in total — as above]
$\tan x$ repeats every π radians, so keep adding π to the original solution to find the other solutions in the interval.

5 Using $\sin^2 x \equiv 1 - \cos^2 x$:

$7 - 3\cos x = 9\sin^2 x \Rightarrow 7 - 3\cos x = 9(1 - \cos^2 x)$ *[1 mark]*

$\Rightarrow 7 - 3\cos x = 9 - 9\cos^2 x$

$\Rightarrow 9\cos^2 x - 3\cos x - 2 = 0$ *[1 mark]*

Factorising the quadratic gives:

$(3\cos x - 2)(3\cos x + 1) = 0$ *[1 mark]*

$\Rightarrow \cos x = \dfrac{2}{3}$ or $\cos x = -\dfrac{1}{3}$ *[1 mark]*

$\cos x = \dfrac{2}{3} \Rightarrow x = 48.189...° = 48.2°$ (1 d.p.)

$\cos x = -\dfrac{1}{3} \Rightarrow x = 109.471...° = 109.5°$ (1 d.p.)

[1 mark for both correct solutions]
[5 marks available in total — as above]

6 $\sin 3\pi t = \dfrac{6\sqrt{2}}{12} = \dfrac{\sqrt{2}}{2}$ *[1 mark]*

$\Rightarrow 3\pi t = \sin^{-1}\left(\dfrac{\sqrt{2}}{2}\right)$ *[1 mark]* $= \dfrac{\pi}{4}, \dfrac{3\pi}{4}, \dfrac{9\pi}{4}, \dfrac{11\pi}{4}...$ *[1 mark]*

So the first three solutions are:

$t = \dfrac{\pi}{4} \div 3\pi = \dfrac{1}{12}$, $t = \dfrac{3\pi}{4} \div 3\pi = \dfrac{1}{4}$, $t = \dfrac{9\pi}{4} \div 3\pi = \dfrac{3}{4}$

[2 marks for all three correct solutions, otherwise 1 mark for any one correct solution]
[5 marks available in total — as above]

Pages 24-25 — Addition and Double Angle Formulas

1 a) $\sin 2x = 2\sin x \cos x = 2 \times \dfrac{2}{3} \times \dfrac{\sqrt{5}}{3} = \dfrac{4\sqrt{5}}{9}$

[2 marks available — 1 mark for using a correct identity, 1 mark for the correct answer]

b) $\cos 2x = \cos^2 x - \sin^2 x = \dfrac{\left(\sqrt{5}\right)^2}{3^2} - \dfrac{2^2}{3^2} = \dfrac{5}{9} - \dfrac{4}{9} = \dfrac{1}{9}$

[2 marks available — 1 mark for using a correct identity, 1 mark for the correct answer]

2 $\dfrac{1 + \cos x}{2} \equiv \dfrac{1}{2}\left(1 + \cos 2\left(\dfrac{x}{2}\right)\right)$

$\equiv \dfrac{1}{2}\left(1 + \left(2\cos^2 \dfrac{x}{2} - 1\right)\right)$

$\equiv \dfrac{1}{2}\left(2\cos^2 \dfrac{x}{2}\right) \equiv \cos^2 \dfrac{x}{2}$

[2 marks available – 1 mark for using a correct identity, 1 mark for the correct rearrangement]

3 a) $\cos\left(x - \dfrac{\pi}{4}\right) = \cos x \cos \dfrac{\pi}{4} + \sin x \sin \dfrac{\pi}{4}$

$= \dfrac{\sqrt{2}}{2}\cos x + \dfrac{\sqrt{2}}{2}\sin x$

$= \dfrac{\sqrt{2}}{2}(\cos x + \sin x)$

So $a = 2$ and $b = 2$.

[2 marks available — 1 mark for a, 1 mark for b]

b) $\dfrac{5\pi}{12} = x - \dfrac{\pi}{4}$, then $x = \dfrac{5\pi}{12} + \dfrac{\pi}{4} = \dfrac{2\pi}{3}$ *[1 mark]*

So $\cos\left(\dfrac{2\pi}{3} - \dfrac{\pi}{4}\right) = \dfrac{\sqrt{2}}{2}\left(\cos\left(\dfrac{2\pi}{3}\right) + \sin\left(\dfrac{2\pi}{3}\right)\right)$

$= \dfrac{\sqrt{2}}{2} \times \left(-\dfrac{1}{2} + \dfrac{\sqrt{3}}{2}\right)$ *[1 mark]*

$= \dfrac{\sqrt{2}}{2} \times \left(\dfrac{-1 + \sqrt{3}}{2}\right) = \dfrac{\sqrt{6} - \sqrt{2}}{4}$ *[1 mark]*

[3 marks available in total — as above]

4 Using $\cos 2x \equiv 2\cos^2 x - 1$:
$\cos 2x + 7\cos x = -4 \Rightarrow 2\cos^2 x - 1 + 7\cos x = -4$ *[1 mark]*

$\Rightarrow 2\cos^2 x + 7\cos x + 3 = 0$ *[1 mark]*

$\Rightarrow (\cos x + 3)(2\cos x + 1) = 0$ *[1 mark]*

$\Rightarrow \cos x = -3$ (no solutions) or $\cos x = -\dfrac{1}{2}$ *[1 mark]*

$\Rightarrow x = 120°$ or $240°$ *[1 mark]*

[5 marks available in total — as above]

5 Use the triangles to find the necessary cos and sin values:

$AB = \sqrt{4^2 + 5^2} = \sqrt{41}$, and $AD = \sqrt{3^2 + 4^2} = 5$ *[1 mark]*

$\sin x = \dfrac{5}{\sqrt{41}} = \dfrac{5\sqrt{41}}{41}$ and $\cos x = \dfrac{4}{\sqrt{41}} = \dfrac{4\sqrt{41}}{41}$ *[1 mark]*

$\sin y = \dfrac{3}{5}$ and $\cos y = \dfrac{4}{5}$ *[1 mark]*

$\sin(x - y) = \sin x \cos y - \cos x \sin y$

$= \dfrac{5\sqrt{41}}{41} \times \dfrac{4}{5} - \dfrac{4\sqrt{41}}{41} \times \dfrac{3}{5}$ *[1 mark]*

$= \dfrac{20\sqrt{41}}{205} - \dfrac{12\sqrt{41}}{205} = \dfrac{8\sqrt{41}}{205}$ *[1 mark]*

[5 marks available in total — as above]

6 $\sin 2\theta \equiv 2\sin\theta\cos\theta$, so $3\sin 2\theta\tan\theta = 6\sin\theta\cos\theta\tan\theta$ *[1 mark]*

As $\tan\theta = \dfrac{\sin\theta}{\cos\theta}$, $6\sin\theta\cos\theta\,\dfrac{\sin\theta}{\cos\theta} = 6\sin^2\theta$ *[1 mark]*

so $3\sin 2\theta\tan\theta = 5 \Rightarrow 6\sin^2\theta = 5$

Then $\sin^2\theta = \dfrac{5}{6} \Rightarrow \sin\theta = \pm\sqrt{\dfrac{5}{6}} = \pm\,0.9128...$ *[1 mark]*

$\sin\theta = 0.9128 \Rightarrow \theta = 1.15,\ 1.99$ (3 s.f.) *[1 mark]*

$\sin\theta = -0.9128 \Rightarrow \theta = 4.29,\ 5.13$ (3 s.f.) *[1 mark]*

[5 marks available in total — as above]

Remember to include the solutions for the negative square root as well. Drawing a sketch here is really useful — you can see that there are 4 solutions you need to find.

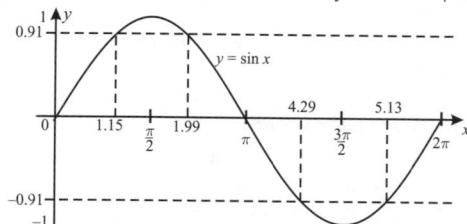

Pages 26-27 — The Wave Function

1 a) $P(\theta) = 7\sqrt{2}\sin\theta + \sqrt{2}\cos\theta \equiv k\cos(\theta - \alpha)$
$\equiv k\cos\theta\cos\alpha + k\sin\theta\sin\alpha$ *[1 mark]*

$\Rightarrow k\sin\alpha = 7\sqrt{2}$ and $k\cos\alpha = \sqrt{2}$ *[1 mark]*

$k = \sqrt{(7\sqrt{2})^2 + (\sqrt{2})^2} = \sqrt{98 + 2} = \sqrt{100} = 10$ *[1 mark]*

Both $\sin\alpha$ and $\cos\alpha$ are positive, so using an ASTC diagram:

$\dfrac{\text{S} \mid \boxed{\text{A}}}{\text{T} \mid \text{C}} \Rightarrow 0 < \alpha < \dfrac{\pi}{2}$. Then $\tan\alpha = \dfrac{7}{1} = 7 \Rightarrow \alpha = 1.428...$

So $P(\theta) \equiv 10\cos(\theta - 1.428...)$ *[1 mark]*

[4 marks available in total — as above]

b) Maximum value of $P(\theta)$ is when $\cos(\theta - 1.428...) = 1$
$\Rightarrow P(\theta) = 10$ *[1 mark]*
Minimum value of $P(\theta)$ is when $\cos(\theta - 1.428...) = -1$
$\Rightarrow P(\theta) = -10$ *[1 mark]*
[2 marks available in total — as above]

2 a) $3\cos 2x - 2\sin 2x \equiv k\sin(2x + \alpha)$
$\equiv k\sin 2x\cos\alpha + k\cos 2x\sin\alpha$ *[1 mark]*

$\Rightarrow k\sin\alpha = 3$ and $k\cos\alpha = -2$ *[1 mark]*

$k = \sqrt{(-2)^2 + (3)^2} = \sqrt{4 + 9} = \sqrt{13}$ *[1 mark]*

$\sin\alpha$ is positive and $\cos\alpha$ is negative, so using an ASTC diagram:

$\dfrac{\boxed{\text{S}} \mid \text{A}}{\text{T} \mid \text{C}} \Rightarrow \dfrac{\pi}{2} < \alpha < \pi$.

Then $\tan\alpha = \dfrac{3}{-2} \Rightarrow \alpha = 2.158...$ *[1 mark]*

[4 marks available in total — as above]

b) $3\cos 2x - 2\sin 2x = 1 \Rightarrow \sqrt{13}\sin(2x + 2.158...) = 1$ *[1 mark]*
$\Rightarrow \sin(2x + 2.158...) = \dfrac{1}{\sqrt{13}}$ *[1 mark]*

$2x + 2.158... = \sin^{-1}\left(\dfrac{1}{\sqrt{13}}\right)$
$= 0.281...,\ 2.860...,\ 6.564...,$ etc. *[1 mark]*

0.281... will be outside of the range, so:
$2x + 2.158... = 2.860...$ and $6.564...$
$\Rightarrow 2x = 0.701...$ and $4.405...$
$\Rightarrow x = 0.3508...$ and $x = 2.202...$

So A is at $x = 0.351$ (3 s.f.) and B is at $x = 2.20$ (3 s.f.) *[1 mark]*

[4 marks available in total — as above]

3 $h(t) = 14 + 3\sin t + 5\cos t \equiv 14 + k\cos(t + \alpha)$
$\equiv 14 + k\cos t\cos\alpha - k\sin t\sin\alpha$ *[1 mark]*

$\Rightarrow k\sin\alpha = -3$ and $k\cos\alpha = 5$ *[1 mark]*

$k = \sqrt{3^2 + 5^2} = \sqrt{9 + 25} = \sqrt{34}$ *[1 mark]*

$\sin\alpha$ is negative and $\cos\alpha$ is positive, so using an ASTC diagram:

$\dfrac{\text{S} \mid \text{A}}{\text{T} \mid \boxed{\text{C}}} \Rightarrow 270° < \alpha < 360°$. Then $\tan\alpha = \dfrac{-3}{5} \Rightarrow \alpha = 329.03...°$

So $h(t) \equiv 14 + \sqrt{34}\cos(t + 329.03...°)$ *[1 mark]*

The maximum of $\cos$ is 1, so the maximum of $h(t)$ is:
$14 + \sqrt{34}\,(1) = 19.8$ m (1 d.p.) *[1 mark]*

This will happen when $\cos(t + 329.03...°) = 1$ *[1 mark]*
$\Rightarrow t + 329.03...° = 0°,\ 360°,\ 720°...$
$\Rightarrow t = -329.03...,\ 30.963...,\ 390.963...$
So the first time that it reaches the maximum height
will be at $t = 31.0$ minutes (1 d.p.) *[1 mark]*

[7 marks available in total — as above]

4 a) $\sqrt{2}\cos\theta - 3\sin\theta \equiv k\sin(\theta - \alpha)$
$\equiv k\sin\theta\cos\alpha - k\cos\theta\sin\alpha$ *[1 mark]*

$\Rightarrow k\sin\alpha = -\sqrt{2}$ and $k\cos\alpha = -3$ *[1 mark]*

$k = \sqrt{(-\sqrt{2})^2 + (-3)^2} = \sqrt{2 + 9} = \sqrt{11}$ *[1 mark]*

Both $\sin\alpha$ and $\cos\alpha$ are negative, so using an ASTC diagram:

$\dfrac{\text{S} \mid \text{A}}{\boxed{\text{T}} \mid \text{C}} \Rightarrow \pi < \alpha < \dfrac{3\pi}{2}$. Then $\tan\alpha = \dfrac{\sqrt{2}}{3} \Rightarrow \alpha = 3.582...$

So $\sqrt{2}\cos\theta - 3\sin\theta = \sqrt{11}\sin(\theta - 3.582...)$ *[1 mark]*

[4 marks available in total — as above]

b) $\sqrt{2}\cos\theta - 3\sin\theta = 3 \Rightarrow \sqrt{11}\sin(\theta - 3.582...) = 3$ *[1 mark]*
$\Rightarrow \sin(\theta - 3.582...) = \dfrac{3}{\sqrt{11}}$ *[1 mark]*

$\Rightarrow \theta - 3.582... = -4.271...,\ 1.130...,\ 2.011...,\ 7.413...,$ etc.
$\Rightarrow \theta = -0.689...,\ 4.712...,\ 5.593...,\ 10.99...,$ etc.

The first two solutions where $\theta > 0$ are:
$\theta = 4.712...$ *[1 mark]* and $\theta = 5.593...$ *[1 mark]*

So the water sprays 3 feet to the right of the sprinkler at
4 mins 43 seconds and 5 mins 36 seconds. *[1 mark for both]*

[5 marks available in total — as above]

Section Three — Geometric Skills

Pages 28-30: Linear Coordinate Geometry

1 As the lines are parallel, the line also has a gradient of $-\dfrac{1}{2}$ *[1 mark]*
Using $y - b = m(x - a)$:
$y - 7 = -\dfrac{1}{2}(x - (-4)) \Rightarrow y - 7 = -\dfrac{x}{2} - 2$
So the equation of the line is $y = -\dfrac{x}{2} + 5$. *[1 mark]*
[2 marks available in total — as above]

2 a) Find the gradient of the line opposite to A, which is BC:
$m = \dfrac{y_2 - y_1}{x_2 - x_1} = \dfrac{(-3) - 3}{5 - (-4)} = -\dfrac{6}{9} = -\dfrac{2}{3}$ *[1 mark]*
So the gradient of the altitude from A is $-1 \div -\dfrac{2}{3} = \dfrac{3}{2}$ *[1 mark]*
Using $y - b = m(x - a)$:
$y - 7 = \dfrac{3}{2}(x - 3) \Rightarrow y - 7 = \dfrac{3x}{2} - \dfrac{9}{2}$
So the equation of the line is $y = \dfrac{3x}{2} + \dfrac{5}{2}$. *[1 mark]*
[3 marks available in total — as above]

b) Find the intersection of the two lines by substitution:
$5\left(\dfrac{3}{2}x + \dfrac{5}{2}\right) - x = 19 \Rightarrow \dfrac{15}{2}x + \dfrac{25}{2} - x = 19$
$\Rightarrow 15x + 25 - 2x = 38$
$\Rightarrow 13x = 13 \Rightarrow x = 1$ *[1 mark]*
Substitute $x = 1$ into $5y - x = 19$:
$5y - 1 = 19 \Rightarrow 5y = 20 \Rightarrow y = 4$
So the orthocentre is at $(1, 4)$ *[1 mark]*
[2 marks available in total — as above]

3   The intersection of line l and BC is the midpoint of BC.

Midpoint $= \left(\dfrac{-1+1}{2}, \dfrac{0+3}{2}\right) = \left(0, \dfrac{3}{2}\right)$ *[1 mark]*

Find the gradient, m, of line l by using A and the midpoint:

$m = \dfrac{\frac{3}{2} - (4)}{0 - (-2)} = -\dfrac{5}{4}$ *[1 mark]*

Using $y - b = m(x - a)$:

$y - 4 = -\dfrac{5}{4}(x - (-2)) \Rightarrow y = -\dfrac{5}{4}x - \dfrac{5}{2} + 4$

$\Rightarrow y = -\dfrac{5}{4}x + \dfrac{3}{2}$ *[1 mark]*

[3 marks available in total — as above]

4   a)   The line l passes through $(1, k)$, so:

$k + (2 \times 1) - 5 = 0 \Rightarrow k = 3$ *[1 mark]*

So A has coordinates $(1, 3)$ *[1 mark]*

$y + 2x - 5 = 0 \Rightarrow y = -2x + 5$, so the gradient of line l is -2.

The line perpendicular to l has gradient $-1 \div -2 = \dfrac{1}{2}$ *[1 mark]*

Using $y - b = m(x - a)$:

$y - 3 = \dfrac{1}{2}(x - 1) \Rightarrow y - 3 = \dfrac{1}{2}x - \dfrac{1}{2}$

So the equation of the line is $y = \dfrac{1}{2}x + \dfrac{5}{2}$. *[1 mark]*

[3 marks available in total — as above]

   b)   The gradient of line l is -2, so $\tan p = -2$ *[1 mark]*

$\tan^{-1}(-2) = -63.4...°$, so:

$\Rightarrow p = -63.4...° + 180° = 117°$ (3 s.f.) *[1 mark]*

[2 marks available in total — as above]

5   Find the gradients of lines DE and EF:

$m_{DE} = \dfrac{1 - (-2)}{(-1) - (-3)} = \dfrac{3}{2}$ *[1 mark]*

$m_{EF} = \dfrac{10 - 1}{5 - (-1)} = \dfrac{9}{6} = \dfrac{3}{2}$ *[1 mark]*

DE and EF have the same gradient and have E as a common point, so points D, E and F are collinear *[1 mark]*

[3 marks available in total — as above]

6   a)   To find the coordinates of A, solve the equations of the lines simultaneously:

l_1: $x - y + 1 = 0 \Rightarrow y = x + 1$

l_2: $2x + y + 8 = 0$

Using substitution:

$2x + (x + 1) + 8 = 0 \Rightarrow 3x = -9 \Rightarrow x = -3$ *[1 mark]*

Put $x = -3$ back into l_1 to find y:

$y = -3 + 1 = -2$

So A has coordinates $(-3, -2)$ *[1 mark]*

[2 marks available in total — as above]

   b)   As D is the midpoint of AC:

$D = \left(\dfrac{(-3) + (-1)}{2}, \dfrac{(-2) + 2}{2}\right) = (-2, 0)$ *[1 mark]*

The gradient (m) of the line through B and D is

$m = \dfrac{0 - (-4)}{(-2) - 6} = \dfrac{4}{-8} = -\dfrac{1}{2}$ *[1 mark]*

Using $y - b = m(x - a)$:

$y - (-4) = -\dfrac{1}{2}(x - 6) \Rightarrow y + 4 = -\dfrac{1}{2}x + 3$

$\Rightarrow y = -\dfrac{1}{2}x - 1$ *[1 mark]*

[3 marks available in total — as above]

   c)   You need to prove that lines AD and BD are perpendicular.

Gradient of AD is:

$m_{AD} = \dfrac{0 - (-2)}{(-2) - (-3)} = 2$ *[1 mark]*

From b), the gradient of BD is $-\dfrac{1}{2}$, so $m_{BD} \times m_{AD} = -\dfrac{1}{2} \times 2 = -1$

So angle ADB is a right angle. *[1 mark]*

[2 marks available in total — as above]

7   a)   The perpendicular bisectors of AC and BC pass through the circumcentre, so use this to calculate the gradient of AC and BC:

Gradient of AC's perpendicular bisector $= \dfrac{2.5 - 3}{3 - 2} = -\dfrac{1}{2}$

$\Rightarrow$ Gradient of AC $= -1 \div -\dfrac{1}{2} = 2$ *[1 mark]*

Using $y - b = m(x - a)$:

$y - 3 = 2(x - 2) \Rightarrow y - 3 = 2x - 4 \Rightarrow y = 2x - 1$ *[1 mark]*

Gradient of BC's perpendicular bisector $= \dfrac{2.5 - 3}{3 - 4} = \dfrac{1}{2}$

$\Rightarrow$ Gradient of BC $= -1 \div \dfrac{1}{2} = -2$ *[1 mark]*

Using $y - b = m(x - a)$:

$y - 3 = -2(x - 4) \Rightarrow y - 3 = -2x + 8 \Rightarrow y = -2x + 11$ *[1 mark]*

Using substitution:

$2x - 1 = -2x + 11 \Rightarrow 4x = 12 \Rightarrow x = 3$ *[1 mark]*

Then $y = 2x - 1 \Rightarrow y = 2(3) - 1 = 5$

So point C has the coordinates $(3, 5)$ *[1 mark]*.

[6 marks available in total — as above]

   b)   The midpoint of AB and the circumcentre have x-values of 3, so the perpendicular bisector of AB is $x = 3$.

Since the perpendicular bisector is vertical,

AB must be a horizontal line through the point $(3, 1)$.

So the equation of AB is $y = 1$.

[2 marks available — 1 mark for a correct method, 1 mark for the correct answer]

Pages 31-34: Circle Geometry

1    $a = 3$, $b = -2$, $r = \sqrt{5}$, so $(x - 3)^2 + (y + 2)^2 = 5$ *[1 mark]*

2   a)   Complete the square for the x and y terms:

$x^2 + y^2 - 4x + 1 = 0$

$\Rightarrow x^2 - 4x + y^2 + 1 = 0$

$\Rightarrow (x - 2)^2 - 4 + y^2 + 1 = 0$

$\Rightarrow (x - 2)^2 + y^2 = 3$

So the coordinates of the centre are $(2, 0)$ *[1 mark]*

   b)   Radius $= \sqrt{3}$ *[1 mark]*

3   a)   The centre of the circle must be the midpoint of AB, since AB is a diameter. Midpoint of AB is:

$\left(\dfrac{2 + 0}{2}, \dfrac{1 + (-5)}{2}\right) = (1, -2)$ *[1 mark]*

The radius is the distance from the centre $(1, -2)$ to point A:

radius $= \sqrt{(2 - 1)^2 + (1 - (-2))^2} = \sqrt{10}$ *[1 mark]*

[2 marks available in total — as above]

You could have used the coordinates of B instead of A here.

   b)   The circle has equation $(x - 1)^2 + (y + 2)^2 = 10$.

To show that the point $(4, -1)$ lies on the circle, show that it satisfies the equation of the circle:

$(4 - 1)^2 + (-1 + 2)^2 = 9 + 1 = 10$,

so $(4, -1)$ lies on the circle.

[2 marks available — 1 mark for a correct method, 1 mark for the correct answer]

   c)   Using the centre of the circle and point A:

Gradient of the radius $= \dfrac{1 - (-2)}{2 - 1} = 3$ *[1 mark]*

The tangent at point A is perpendicular to the radius at point A, so the tangent has gradient $-1 \div 3 = -\dfrac{1}{3}$ *[1 mark]*.

Using $y - b = m(x - a)$:

$y - 1 = -\dfrac{1}{3}(x - 2) \Rightarrow y - 1 = -\dfrac{1}{3}x + \dfrac{2}{3}$

$\Rightarrow y = -\dfrac{1}{3}x + \dfrac{5}{3}$ *[1 mark]*

[3 marks available in total — as above]

4 a) The line through the centre P bisects the chord, and so is perpendicular to the chord AB at the midpoint M.

Gradient of AB = Gradient of AM = $\frac{(7-10)}{(11-9)} = -\frac{3}{2}$ *[1 mark]*

Gradient of PM = $-1 \div -\frac{3}{2} = \frac{2}{3} = \frac{(7-3)}{(11-p)}$ *[1 mark]*

$\Rightarrow 3(7-3) = 2(11-p) \Rightarrow 12 = 22 - 2p \Rightarrow p = 5$ *[1 mark]*

[3 marks available in total — as above]

b) r is the distance from P to A:

$r = \sqrt{(9-5)^2 + (10-3)^2} = \sqrt{4^2 + 7^2} = \sqrt{65}$

$\Rightarrow r^2 = 65$ *[1 mark]*

So the equation of the circle is:

$(x-5)^2 + (y-3)^2 = 65$ *[1 mark]*

[2 marks available in total — as above]

5 a) The radius of the larger circle is the distance from A to $(-5, -4)$:

$r = \sqrt{((-3)-(-5))^2 + (6-(-4))^2} = \sqrt{2^2 + 10^2} = \sqrt{104}$ *[1 mark]*

Then the distance from B to the centre is:

$\sqrt{(b-(-5))^2 + ((-6)-(-4))^2} = \sqrt{104}$ *[1 mark]*

$\Rightarrow (b+5)^2 + (-2)^2 = 104 \Rightarrow (b+5)^2 = 100$ *[1 mark]*

$\Rightarrow b + 5 = 10$ or $b + 5 = -10$

$\Rightarrow b = 5$ or $b = -15$

b is negative, so $b = -15$ *[1 mark]*

[4 marks available in total — as above]

b) The midpoint of the chord AB is the point where it touches the smaller circle:

Midpoint of AB = $\left(\frac{(-15)+(-3)}{2}, \frac{6+(-6)}{2}\right) = (-9, 0)$ *[1 mark]*

The distance between $(-9, 0)$ and the centre $(-5, -4)$ is equal to the radius of the smaller circle, so:

$r = \sqrt{((-9)-(-5))^2 + (0-(-4))^2} = \sqrt{(-4)^2 + (-4)^2} = \sqrt{32}$

[1 mark]

So the equation of the circle is: $(x+5)^2 + (y+4)^2 = 32$ *[1 mark]*

[3 marks available in total — as above]

6 a) Substitute $x = 4$ and $y = 3$ into the equation for C:

$(4+2)^2 + (3-1)^2 = 6^2 + 2^2 = 36 + 4 = 40$ *[1 mark]*

b) The centre of C is $(-2, 1)$. A line between the centre of C and A forms the radius with a gradient of $\frac{3-1}{4-(-2)} = \frac{1}{3}$. *[1 mark]*

The tangent at $(4, 3)$ is perpendicular to the radius at $(4, 3)$, so the tangent has gradient $-1 \div \frac{1}{3} = -3$. *[1 mark]*

Using $y - b = m(x - a)$:

$y - 3 = -3(x - 4) \Rightarrow y - 3 = -3x + 12$

$\Rightarrow y = -3x + 15$ *[1 mark]*

[3 marks available in total — as above]

7 a) Change the equation of C_1 into the other form:

$x^2 + y^2 + 6x - 4y = 23$

$\Rightarrow (x+3)^2 - 9 + (y-2)^2 - 4 = 23$

$\Rightarrow (x+3)^2 + (y-2)^2 = 36$ *[1 mark]*

So C_1 has centre $(-3, 2)$ *[1 mark]*

Using Pythagoras' theorem, the distance between the centres of the two circles, d, is given by:

$d = \sqrt{((-6)-2)^2 + (k-(-3))^2}$

$\Rightarrow d^2 = 64 + (k+3)^2$ *[1 mark]*

The distance between the centres of the two circles is equal to the sum of the two radii, so $d = r_1 + r_2 = 6 + 4 = 10$ *[1 mark]*

$\Rightarrow 10^2 = 64 + (k+3)^2$ *[1 mark]*

$\Rightarrow (k+3)^2 = 100 - 64 \Rightarrow (k+3)^2 = 36$

$\Rightarrow k + 3 = 6$ or $k + 3 = -6 \Rightarrow k = 3$ or $k = -9$

k is positive so $k = 3$ *[1 mark]*

[6 marks available in total — as above]

b) $(x-3)^2 + (y+6)^2 = 16$ *[1 mark]*

8 a) $x^2 + y^2 + 10x - 6y + 25 = 0$

$\Rightarrow (x+5)^2 - 25 + (y-3)^2 - 9 + 25 = 0$

$\Rightarrow (x+5)^2 + (y-3)^2 = 9$

So C_1 has centre $(-5, 3)$ and a radius of 3. *[1 mark]*

Using the ratios, C_2 has a radius of 6. *[1 mark]*

Then the centre of C_2 is $3 + 6 = 9$ units to the right of the centre of $C_1 \Rightarrow$ the centre of C_2 is $(4, 3)$ *[1 mark]*

So the equation of C_2 is: $(x-4)^2 + (y-3)^2 = 36$ *[1 mark]*

[4 marks available in total— as above]

b) Using the ratios, C_3 has a radius of 15. *[1 mark]*

The centre of C_3 is $3 + 15 = 18$ units to the right of the centre of $C_1 \Rightarrow$ the centre of C_3 is $(13, 3)$. *[1 mark]*

Point B is 15 units above the centre of C_3, so the coordinates of point B are $(13, 18)$ *[1 mark]*

[3 marks available in total — as above]

9 Centre of $C_1 = (4, 3)$ *[1 mark]*

Radius of $C_1 = \sqrt{8}$ *[1 mark]*

$x^2 + y^2 - 2x + 4y - 3 = 0$

$\Rightarrow (x-1)^2 - 1 + (y+2)^2 - 4 - 3 = 0$

$\Rightarrow (x-1)^2 + (y+2)^2 = 8$

Centre of $C_2 = (1, -2)$ *[1 mark]*

Radius of $C_2 = \sqrt{8}$ *[1 mark]*

Sum of the radii = $\sqrt{8} + \sqrt{8} = 2\sqrt{8} = \sqrt{32}$ *[1 mark]*

Distance between the centres = $\sqrt{(-2-3)^2 + (1-4)^2}$

$= \sqrt{(-5)^2 + (-3)^2}$

$= \sqrt{34}$ *[1 mark]*

$\sqrt{34} > \sqrt{32}$, so the distance between the centres of the circles is more than the sum of their radii. Therefore, the circles do not intersect. *[1 mark]*

[7 marks available in total — as above]

Pages 35-36: Solving Geometrical Problems

1 a) Rearrange the second equation: $x = 3y - 5$

Substitute $x = 3y - 5$ into $x^2 + y^2 = 5$:

$(3y-5)^2 + y^2 = 5$ *[1 mark]*

$\Rightarrow 9y^2 - 30y + 25 + y^2 - 5 = 0$

$\Rightarrow 10y^2 - 30y + 20 = 0$

$\Rightarrow y^2 - 3y + 2 = 0$ as required *[1 mark]*

[2 marks available in total — as above]

b) $y^2 - 3y + 2 = 0 \Rightarrow (y-2)(y-1) = 0$ *[1 mark]*

$\Rightarrow y = 2$ or $y = 1$ *[1 mark]*

When $y = 2$, $x = 3(2) - 5 = 1$

When $y = 1$, $x = 3(1) - 5 = -2$

So the circle and line intersect at $(1, 2)$ and $(-2, 1)$. *[1 mark]*

[3 marks available in total — as above]

2 Set the two equations equal to each other:

$x^3 + x^2 - 3x + 4 = 3x + 4$ *[1 mark]*

$\Rightarrow x^3 + x^2 - 3x - 3x - 4 = 0 \Rightarrow x^3 + x^2 - 6x = 0$ *[1 mark]*

$\Rightarrow x(x^2 + x - 6) = 0 \Rightarrow x(x+3)(x-2) = 0$ *[1 mark]*

$\Rightarrow x = -3, 0$ or 2 *[1 mark]*

When $x = -3$, $y = 3(-3) + 4 = -5$

When $x = 0$, $y = 3(0) + 4 = 4$

When $x = 2$, $y = 3(2) + 4 = 10$

So the points of intersection are $(-3, -5)$, $(0, 4)$ and $(2, 10)$ *[1 mark]*

[5 marks available in total — as above]

3 Rearrange $x + y = 4$ into $y = -x + 4$, and substitute it into the circle equation:

$(x+2)^2 + (y-8)^2 = 52 \Rightarrow (x+2)^2 + (-x+4-8)^2 = 52$ *[1 mark]*

$\Rightarrow x^2 + 4x + 4 + (-x-4)^2 = 52$

$\Rightarrow x^2 + 4x + 4 + x^2 + 8x + 16 - 52 = 0$

$\Rightarrow 2x^2 + 12x - 32 = 0 \Rightarrow x^2 + 6x - 16 = 0$ *[1 mark]*

$\Rightarrow (x+8)(x-2) = 0$ *[1 mark]*

$\Rightarrow x = -8$ or $x = 2$ *[1 mark]*

When $x = -8$, $y = -(-8) + 4 = 12$

When $x = 2$, $y = -2 + 4 = 2$

So the points of intersection are $(-8, 12)$ and $(2, 2)$ *[1 mark]*

[5 marks available in total — as above]

4   Rearrange $2y - x + 2 = 0$ into $x = 2y + 2$,
    and substitute it into the circle equation:
    $(x + 7)^2 + (y + 2)^2 = 5 \Rightarrow (2y + 2 + 7)^2 + (y + 2)^2 = 5$ *[1 mark]*
    $\Rightarrow (2y + 9)^2 + y^2 + 4y + 4 = 5$
    $\Rightarrow 4y^2 + 36y + 81 + y^2 + 4y + 4 - 5 = 0$
    $\Rightarrow 5y^2 + 40y + 80 = 0 \Rightarrow y^2 + 8y + 16 = 0$ *[1 mark]*
    $\Rightarrow (y + 4)^2 = 0 \Rightarrow y = -4$ *[1 mark]*
    There is only one solution, so the line touches the circle in one place
    and must be a tangent to the circle. *[1 mark]*
    Substitute $y = -4$ into $x = 2y + 2$: $x = 2(-4) + 2 = -6$,
    so the coordinates of the point of intersection are $(-6, -4)$. *[1 mark]*
    [5 marks available in total — as above]
    You could also have rearranged the equation of the line to $y = 0.5x - 1$
    and substituted that into the equation of the circle instead, but using the
    method above means you only have to deal with whole numbers.

5   Set the two equations equal to each other:
    $x^3 + 5x^2 - 2x - 15 = x^3 - x^2 + 4x - 3$ *[1 mark]*
    $\Rightarrow x^3 - x^3 + 5x^2 + x^2 - 2x - 4x - 15 + 3 = 0$
    $\Rightarrow 6x^2 - 6x - 12 = 0 \Rightarrow x^2 - x - 2 = 0$ *[1 mark]*
    $\Rightarrow (x + 1)(x - 2) = 0$ *[1 mark]*
    $\Rightarrow x = -1$ or $x = 2$ *[1 mark]*
    When $x = -1$, $y = (-1)^3 + 5(-1)^2 - 2(-1) - 15 = -9$
    When $x = 2$, $y = (2)^3 + 5(2)^2 - 2(2) - 15 = 9$
    So the points of intersection are $(-1, -9)$ and $(2, 9)$ *[1 mark]*
    [5 marks available in total — as above]

6   Rearrange $2x - y - 1 = 0$ into $y = 2x - 1$,
    and substitute it into the circle equation:
    $(x + 3)^2 + (y - 3)^2 = 15 \Rightarrow (x + 3)^2 + (2x - 1 - 3)^2 = 15$ *[1 mark]*
    $\Rightarrow x^2 + 6x + 9 + (2x - 4)^2 = 15$
    $\Rightarrow x^2 + 6x + 9 + 4x^2 - 16x + 16 - 15 = 0$
    $\Rightarrow 5x^2 - 10x + 10 = 0$
    $\Rightarrow x^2 - 2x + 2 = 0$ *[1 mark]*
    Find the discriminant of this quadratic equation:
    $b^2 - 4ac = (-2)^2 - (4 \times 1 \times 2) = -4$ *[1 mark]*
    The discriminant is negative, so the quadratic has no solutions.
    Therefore, the line and circle do not intersect. *[1 mark]*
    [4 marks available in total — as above]

Pages 37-39: Vectors

1 a) $3\mathbf{a} - 2\mathbf{b} = 3\begin{pmatrix} 2 \\ 3 \\ -2 \end{pmatrix} - 2\begin{pmatrix} 1 \\ 4 \\ 1 \end{pmatrix}$

       $= \begin{pmatrix} 6 \\ 9 \\ -6 \end{pmatrix} - \begin{pmatrix} 2 \\ 8 \\ 2 \end{pmatrix} = \begin{pmatrix} 4 \\ 1 \\ -8 \end{pmatrix}$ *[1 mark]*

   b) $|3\mathbf{a} - 2\mathbf{b}| = \sqrt{4^2 + 1^2 + (-8)^2} = \sqrt{81} = 9$
       [2 marks available — 1 mark for a correct method,
       1 mark for the correct answer]

2   The magnitude of $\mathbf{a}$ is:
    $\sqrt{4^2 + (-4)^2 + 7^2} = \sqrt{16 + 16 + 49} = \sqrt{81} = 9$
    So the unit vector in the direction of $\mathbf{a}$ is:

    $\frac{1}{|\mathbf{a}|} \times \mathbf{a} = \frac{1}{9}\begin{pmatrix} 4 \\ -4 \\ 7 \end{pmatrix} = \begin{pmatrix} \frac{4}{9} \\ -\frac{4}{9} \\ \frac{7}{9} \end{pmatrix}$

    [3 marks available — 1 mark for a correct method, 1 mark
    for the correct magnitude, 1 mark for the correct answer]

3 a) M is the midpoint of $AC \Rightarrow \overrightarrow{AM} = \frac{1}{2}\overrightarrow{AC}$ *[1 mark]*
                          $= \frac{1}{2}(-\mathbf{s} + \mathbf{r})$ *[1 mark]*
       [2 marks available in total — as above]

   b) E.g. $\overrightarrow{ME} = \overrightarrow{MA} + \overrightarrow{AD} + \overrightarrow{DE}$
           $= -\frac{1}{2}(-\mathbf{s} + \mathbf{r}) + (-\mathbf{s}) + \mathbf{t}$
           $= \frac{1}{2}(\mathbf{s} - \mathbf{r}) + (-\mathbf{s}) + \mathbf{t} = -\frac{1}{2}(\mathbf{s} + \mathbf{r}) + \mathbf{t}$
       [2 marks available — 1 mark for an appropriate pathway,
       1 mark for the correct answer]

4   $\overrightarrow{AB} = \overrightarrow{OB} - \overrightarrow{OA} = (\mathbf{i} - 7\mathbf{j} + 10\mathbf{k}) - (-5\mathbf{i} + 3\mathbf{j} + 2\mathbf{k})$
        $= 6\mathbf{i} - 10\mathbf{j} + 8\mathbf{k}$ *[1 mark]*
    $\overrightarrow{AM} = \frac{1}{2}\overrightarrow{AB} = 3\mathbf{i} - 5\mathbf{j} + 4\mathbf{k}$
    Use this to find $\overrightarrow{CM}$:
    $\overrightarrow{CM} = -\overrightarrow{OC} + \overrightarrow{OA} + \overrightarrow{AM}$
        $= -(3\mathbf{i} + 2\mathbf{j} + 3\mathbf{k}) + (-5\mathbf{i} + 3\mathbf{j} + 2\mathbf{k}) + (3\mathbf{i} - 5\mathbf{j} + 4\mathbf{k})$
        $= -5\mathbf{i} - 4\mathbf{j} + 3\mathbf{k}$
    [1 mark for a correct method, 1 mark for the correct vector]
    $|\overrightarrow{CM}| = \sqrt{(-5)^2 + (-4)^2 + 3^2} = \sqrt{50}$
    $|\overrightarrow{AB}| = \sqrt{6^2 + (-10)^2 + 8^2} = \sqrt{200}$ *[1 mark for both]*
    $k = \frac{|\overrightarrow{AB}|}{|\overrightarrow{CM}|} = \frac{\sqrt{200}}{\sqrt{50}} = \sqrt{\frac{200}{50}} = \sqrt{4} = 2$ *[1 mark]*
    [5 marks available in total — as above]

5   $\overrightarrow{AB} = \overrightarrow{OB} - \overrightarrow{OA}$
        $= (14\mathbf{i} + 12\mathbf{j} - 9\mathbf{k}) - (-2\mathbf{i} + 4\mathbf{j} - 5\mathbf{k}) = (16\mathbf{i} + 8\mathbf{j} - 4\mathbf{k})$ *[1 mark]*
    $\overrightarrow{AC} = \overrightarrow{OC} - \overrightarrow{OA}$
        $= (2\mathbf{i} + \mu\mathbf{j} + \lambda\mathbf{k}) - (-2\mathbf{i} + 4\mathbf{j} - 5\mathbf{k}) = 4\mathbf{i} + (\mu - 4)\mathbf{j} + (\lambda + 5)\mathbf{k}$
    [1 mark]
    $\overrightarrow{AB}$ and $\overrightarrow{AC}$ both share the point A, so if the points are collinear then
    these vectors are parallel — i.e. $\overrightarrow{AB} = k\overrightarrow{AC}$ for some constant k.
    So: $(16\mathbf{i} + 8\mathbf{j} - 4\mathbf{k}) = k(4\mathbf{i} + (\mu - 4)\mathbf{j} + (\lambda + 5)\mathbf{k})$ *[1 mark]*
    Equate $\mathbf{i}$, $\mathbf{j}$, and $\mathbf{k}$ components:
    $16 = 4k \Rightarrow k = 4$ *[1 mark]*
    $8 = k(\mu - 4) \Rightarrow 8 = 4\mu - 16 \Rightarrow \mu = 6$
    $-4 = k(\lambda + 5) \Rightarrow -4 = 4\lambda + 20 \Rightarrow \lambda = -6$ *[1 mark for both]*
    [5 marks available in total — as above]
    You could have also used that $\overrightarrow{BC}$ is parallel $\overrightarrow{AB}$ or $\overrightarrow{AC}$.

6   $\mathbf{a} - 3\mathbf{b} = (2\mathbf{i} + \mathbf{j} + 3\mathbf{k}) - 3(4\mathbf{i} + m\mathbf{j} - \mathbf{k})$
        $= (2\mathbf{i} + \mathbf{j} + 3\mathbf{k}) - (12\mathbf{i} + 3m\mathbf{j} - 3\mathbf{k})$
        $= (-10\mathbf{i} + (1 - 3m)\mathbf{j} + 6\mathbf{k})$ *[1 mark]*
    $|\mathbf{a} - 3\mathbf{b}| = \sqrt{152} \Rightarrow \sqrt{(-10)^2 + (1 - 3m)^2 + 6^2} = \sqrt{152}$ *[1 mark]*
    $\Rightarrow 100 + (1 - 3m)^2 + 36 = 152$
    $\Rightarrow (1 - 3m)^2 = 16$
    $\Rightarrow 1 - 3m = 4$ or $1 - 3m = -4$
    $\Rightarrow m = -1$ or $m = \frac{5}{3}$ *[1 mark]*
    m is an integer, so $m = -1$. *[1 mark]*
    [4 marks available in total — as above]

7   $\overrightarrow{AB} = \overrightarrow{OB} - \overrightarrow{OA} = \begin{pmatrix} 4 \\ -9 \\ 8 \end{pmatrix} - \begin{pmatrix} 1 \\ -3 \\ 2 \end{pmatrix} = \begin{pmatrix} 3 \\ -6 \\ 6 \end{pmatrix}$

    D is $\frac{2}{3}$ of the way along $\overrightarrow{AB}$, so $\overrightarrow{AD} = \frac{2}{3}\overrightarrow{AB}$ *[1 mark]*

    $\Rightarrow \overrightarrow{AD} = \frac{2}{3}\begin{pmatrix} 3 \\ -6 \\ 6 \end{pmatrix} = \begin{pmatrix} 2 \\ -4 \\ 4 \end{pmatrix}$

    $\overrightarrow{OD} = \overrightarrow{OA} + \overrightarrow{AD} = \begin{pmatrix} 1 \\ -3 \\ 2 \end{pmatrix} + \begin{pmatrix} 2 \\ -4 \\ 4 \end{pmatrix} = \begin{pmatrix} 3 \\ -7 \\ 6 \end{pmatrix}$ *[1 mark]*

    Now, $\overrightarrow{OD} = -\frac{1}{2}\overrightarrow{CE} \Rightarrow \overrightarrow{CE} = -2\overrightarrow{OD}$

    $\Rightarrow \overrightarrow{CE} = -2\begin{pmatrix} 3 \\ -7 \\ 6 \end{pmatrix} = \begin{pmatrix} -6 \\ 14 \\ -12 \end{pmatrix}$ *[1 mark]*

    So $\overrightarrow{OE} = \overrightarrow{OC} + \overrightarrow{CE} = \begin{pmatrix} -3 \\ 9 \\ -4 \end{pmatrix} + \begin{pmatrix} -6 \\ 14 \\ -12 \end{pmatrix} = \begin{pmatrix} -9 \\ 23 \\ -16 \end{pmatrix}$ *[1 mark]*

    [4 marks available in total — as above]

8 a) All sides of the cube have the same length, which means the magnitude of the vectors **a** and **b** are the same.
$\Rightarrow \sqrt{(-7)^2 + (-4)^2 + 4^2} = \sqrt{m^2 + 8^2 + 1^2}$
$\Rightarrow 81 = m^2 + 65 \Rightarrow m^2 = 16 \Rightarrow m = -4$ as m is negative.
[2 marks available — 1 mark for a correct method, 1 mark for the correct answer]

b) $\overrightarrow{OR} = \overrightarrow{OQ} + \overrightarrow{QR}$, so if R has position vector $(-3\mathbf{i} - 3\mathbf{j} + 12\mathbf{k})$, then:
$(-3\mathbf{i} - 3\mathbf{j} + 12\mathbf{k}) = \mathbf{a} + \overrightarrow{QR}$
$\Rightarrow \overrightarrow{QR} = (-3\mathbf{i} - 3\mathbf{j} + 12\mathbf{k}) - (-7\mathbf{i} - 4\mathbf{j} + 4\mathbf{k})$
$= (4\mathbf{i} + \mathbf{j} + 8\mathbf{k})$ *[1 mark]*
As the shape is a cube, $\overrightarrow{QR}$ is parallel to $\overrightarrow{TS}$
$\Rightarrow \overrightarrow{TS} = (4\mathbf{i} + \mathbf{j} + 8\mathbf{k})$
L is the midpoint of TS, so $\overrightarrow{TL} = (2\mathbf{i} + 0.5\mathbf{j} + 4\mathbf{k})$ *[1 mark]*
Then $\overrightarrow{OL} = \mathbf{b} + \overrightarrow{TL} = (-4\mathbf{i} + 8\mathbf{j} + \mathbf{k}) + (2\mathbf{i} + 0.5\mathbf{j} + 4\mathbf{k})$
$= (-2\mathbf{i} + 8.5\mathbf{j} + 5\mathbf{k})$
So $\overrightarrow{RL} = \overrightarrow{OL} - \overrightarrow{OR}$
$= (-2\mathbf{i} + 8.5\mathbf{j} + 5\mathbf{k}) - (-3\mathbf{i} - 3\mathbf{j} + 12\mathbf{k})$
$= (\mathbf{i} + 11.5\mathbf{j} - 7\mathbf{k})$ *[1 mark]*
$|\overrightarrow{RL}| = \sqrt{1^2 + 11.5^2 + (-7)^2} = \sqrt{182.25} = 13.5$ *[1 mark]*
[4 marks available in total — as above]

Pages 40-41: The Scalar Product

1 a) $\mathbf{a}.\mathbf{b} = a_1 b_1 + a_2 b_2 + a_3 b_3 = (-3 \times 4) + (5 \times 3) + (-2 \times -6)$
$= -12 + 15 + 12 = 15$ *[1 mark]*

b) Calculate $|\mathbf{a}|$ and $|\mathbf{b}|$:
$|\mathbf{a}| = \sqrt{(-3)^2 + 5^2 + (-2)^2} = \sqrt{9 + 25 + 4} = \sqrt{38}$ *[1 mark]*
$|\mathbf{b}| = \sqrt{4^2 + 3^2 + (-6)^2} = \sqrt{16 + 9 + 36} = \sqrt{61}$ *[1 mark]*
$\cos\theta = \dfrac{\mathbf{a}.\mathbf{b}}{|\mathbf{a}||\mathbf{b}|} = \dfrac{15}{\sqrt{38} \times \sqrt{61}} = 0.31155...$ *[1 mark]*
$\Rightarrow \theta = \cos^{-1}(0.31155...)$
$= 71.847...° = 71.8°$ (1 d.p.) *[1 mark]*
[4 marks available in total — as above]

2 a) If **s** and **t** are perpendicular, $\mathbf{s}.\mathbf{t} = 0$, so:
$(5 \times 6) + (u \times 2) + (-8 \times 1) = 0$ *[1 mark]*
$\Rightarrow 30 + 2u - 8 = 0 \Rightarrow 2u = -22 \Rightarrow u = -11$ *[1 mark]*
[2 marks available in total — as above]

b) $\mathbf{s}.(\mathbf{s} + \mathbf{t}) = \mathbf{s}.\mathbf{s} + \mathbf{s}.\mathbf{t}$ *[1 mark]*
$\mathbf{s}.\mathbf{t} = 0$, so calculate $\mathbf{s}.\mathbf{s}$:
$(5 \times 5) + (-11 \times -11) + (-8 \times -8)$
$= 25 + 121 + 64 = 210$ *[1 mark]*
[3 marks available in total — as above]

3 $\overrightarrow{PR} = \overrightarrow{PQ} + \overrightarrow{QR} = \begin{pmatrix} 2 \\ -9 \\ 3 \end{pmatrix} + \begin{pmatrix} 14 \\ 6 \\ 7 \end{pmatrix} = \begin{pmatrix} 16 \\ -3 \\ 10 \end{pmatrix}$ *[1 mark]*

$|\overrightarrow{PQ}| = \sqrt{2^2 + (-9)^2 + 3^2} = \sqrt{94}$ *[1 mark]*
$|\overrightarrow{PR}| = \sqrt{16^2 + (-3)^2 + 10^2} = \sqrt{365}$ *[1 mark]*
$\overrightarrow{PQ}.\overrightarrow{PR} = (2 \times 16) + (-9 \times -3) + (3 \times 10)$
$= 32 + 27 + 30 = 89$ *[1 mark]*
$\cos QPR = \dfrac{89}{\sqrt{94} \times \sqrt{365}} = 0.48048...$ *[1 mark]*
$\Rightarrow$ Angle $QPR = \cos^{-1}(0.48048...)$
$= 61.282...° = 61.3°$ (1 d.p.) *[1 mark]*
[6 marks available in total — as above]

4 a) $\overrightarrow{AC} = \overrightarrow{OC} - \overrightarrow{OA} = \begin{pmatrix} 5 \\ -6 \\ 0 \end{pmatrix} - \begin{pmatrix} 3 \\ -2 \\ 6 \end{pmatrix} = \begin{pmatrix} 2 \\ -4 \\ -6 \end{pmatrix}$

$\overrightarrow{OD} = \overrightarrow{OA} + \dfrac{1}{2}\overrightarrow{AC} = \begin{pmatrix} 3 \\ -2 \\ 6 \end{pmatrix} + \dfrac{1}{2}\begin{pmatrix} 2 \\ -4 \\ -6 \end{pmatrix}$
$= \begin{pmatrix} 3 \\ -2 \\ 6 \end{pmatrix} + \begin{pmatrix} 1 \\ -2 \\ -3 \end{pmatrix} = \begin{pmatrix} 4 \\ -4 \\ 3 \end{pmatrix}$
[2 marks available — 1 mark for a correct method, 1 mark for the correct answer]

b) $\overrightarrow{OD}.\overrightarrow{OB} = (4 \times 6) + (-4 \times 0) + (3 \times 0) = 24$ *[1 mark]*
$|\overrightarrow{OD}| = \sqrt{4^2 + (-4)^2 + 3^2} = \sqrt{41}$ *[1 mark]*
$|\overrightarrow{OB}| = \sqrt{6^2 + 0^2 + 0^2} = 6$ *[1 mark]*
$\cos DOB = \dfrac{\overrightarrow{OD}.\overrightarrow{OB}}{|\overrightarrow{OD}||\overrightarrow{OB}|}$ *[1 mark]*
$\Rightarrow \cos DOB = \dfrac{24}{6\sqrt{41}} = 0.62...$
$\Rightarrow$ Angle $DOB = 51.34...° = 51.3°$ (3 s.f.) *[1 mark]*
[5 marks available in total — as above]

5 $\overrightarrow{AE} = \overrightarrow{AD} + \overrightarrow{DE} = \mathbf{s} + \mathbf{t}$
$\overrightarrow{AB}.\overrightarrow{AE} = 32 \Rightarrow \mathbf{r}.(\mathbf{s} + \mathbf{t}) = 32$
So $\mathbf{r}.\mathbf{s} + \mathbf{r}.\mathbf{t} = 32$ *[1 mark]*
r and **s** are perpendicular, so $\mathbf{r}.\mathbf{s} = 0$ *[1 mark]*
Since $ABCD$ is a square, $\overrightarrow{DC} = \overrightarrow{AB} = \mathbf{r}$
So vectors **r** and **t** form an angle of 60°
$\Rightarrow \mathbf{r}.\mathbf{t} = |\mathbf{r}||\mathbf{t}|\cos 60° = \dfrac{1}{2}|\mathbf{r}||\mathbf{t}|$ *[1 mark]*
DCE is an equilateral triangle so $|\mathbf{r}| = |\mathbf{t}|$
$\Rightarrow \mathbf{r}.\mathbf{t} = \dfrac{1}{2}|\mathbf{t}|^2$ *[1 mark]*
So $\mathbf{r}.\mathbf{s} + \mathbf{r}.\mathbf{t} = 32 \Rightarrow 0 + \dfrac{1}{2}|\mathbf{t}|^2 = 32 \Rightarrow |\mathbf{t}| = \sqrt{64} = 8$ *[1 mark]*
[5 marks available in total — as above]

Section Four — Calculus Skills

Pages 42-44 — Differentiation

1 $y = x^7 + 2x^{-3}$
$\dfrac{dy}{dx} = 7x^6 + (-3 \times 2x^{-4}) = 7x^6 - 6x^{-4}$
[3 marks available — 1 mark for writing in a differentiable form, 1 mark for each term correctly differentiated]

2 $\dfrac{dy}{dx} = 3 \times 2\cos 2x = 6\cos 2x$
[2 marks available — 1 mark for differentiating, 1 mark for the correct derivative]

3 To find the gradient of the curve, differentiate $y = 3x + 4 + x^4$:
$\dfrac{dy}{dx} = 3 + 4x^3$ *[1 mark for each correct term]*
To find the gradient at point A, substitute $x = 2$ into the derivative:
gradient $= 3 + 4 \times 2^3 = 35$ *[1 mark]*
[3 marks available in total — as above]

4 $y = x^5 - 4x^3 + x^{-1}$
$\dfrac{dy}{dx} = 5x^4 - 12x^2 - x^{-2}$
At $x = a$, $\dfrac{dy}{dx} = 5a^4 - 12a^2 - \dfrac{1}{a^2}$
At $x = -a$, $\dfrac{dy}{dx} = 5(-a)^4 - 12(-a)^2 - \dfrac{1}{(-a)^2} = 5a^4 - 12a^2 - \dfrac{1}{a^2}$
The gradient of the curve is the same at $x = a$ and $x = -a$, therefore the gradients of the tangents are the same at $x = a$ and $x = -a$.
Lines with the same gradient are parallel, so the tangents at $x = a$ and $x = -a$ are parallel for all values of a.
[4 marks available — 1 mark for writing in a differentiable form, 1 mark for any two terms correctly differentiated, 1 mark for the correct derivative, 1 mark for correct conclusion with justification]

5   Let $u = \sin x$ and $y = 2u^3$, then $\frac{du}{dx} = \cos x$ and $\frac{dy}{du} = 6u^2$

Using the chain rule: $\frac{dy}{dx} = \frac{dy}{du} \times \frac{du}{dx} = 6u^2 \times \cos x = 6(\sin x)^2 \times \cos x$

$\Rightarrow \text{f}'(x) = 6 \times \sin^2(x) \times \cos(x)$

So $\text{f}'\left(\frac{\pi}{3}\right) = 6 \times \sin^2\left(\frac{\pi}{3}\right) \times \cos\left(\frac{\pi}{3}\right) = 6 \times \left(\frac{\sqrt{3}}{2}\right)^2 \times \frac{1}{2} = 3 \times \frac{3}{4} = \frac{9}{4}$

[3 marks available — 1 mark for differentiating, 1 mark for the correct derivative, 1 mark for correct evaluation]

6   $y = 2x^3 - 10x^2 - 4x^{\frac{1}{2}} + 12$

$\frac{dy}{dx} = 6x^2 - 20x - 2x^{-\frac{1}{2}}$

At $x = 4$, $\frac{dy}{dx} = 6 \times 4^2 - 20 \times 4 - \frac{2}{\sqrt{4}} = 96 - 80 - 1 = 15$

[4 marks available — 1 mark for writing in a differentiable form, 1 mark for two terms correctly differentiated, 1 mark for the correct derivative, 1 mark for the correct gradient]

7 a) $\frac{dy}{dx} = 3 \times 6 \times (6x + 4)^2 = 18(6x + 4)^2$

[1 mark for differentiating, 1 mark for the correct derivative]

At $x = -1$, $\frac{dy}{dx} = 18 \times (-6 + 4)^2 = 18 \times 4 = 72$ *[1 mark]*

[3 marks available in total — as above]

   b) $y = \frac{1}{\sqrt{2x - x^2}} = (2x - x^2)^{-\frac{1}{2}}$

Let $u = 2x - x^2$, so $y = u^{-\frac{1}{2}}$, then $\frac{du}{dx} = 2 - 2x$ and $\frac{dy}{du} = -\frac{1}{2}u^{-\frac{3}{2}}$

Using the chain rule: $\frac{dy}{dx} = \frac{dy}{du} \times \frac{du}{dx} = -\frac{1}{2}u^{-\frac{3}{2}} \times (2 - 2x)$

$= -\frac{1}{2}(2x - x^2)^{-\frac{3}{2}} \times (2 - 2x)$

At $x = 1$, $\frac{dy}{dx} = -\frac{1}{2}(2 - 1)^{-\frac{3}{2}} \times (2 - 2) = 0$

[4 marks available — 1 mark for writing in a differentiable form, 1 mark for differentiating, 1 mark for the correct derivative, 1 mark for the correct answer]

8   $\frac{dy}{dx} = 7 \times -2 \times (3 - 2x)^6 = -14(3 - 2x)^6$

[1 mark for differentiating, 1 mark for the correct derivative]

At $x = 2$, $\frac{dy}{dx} = -14 \times (3 - 2 \times 2)^6 = -14 \times (-1)^6 = -14$ *[1 mark]*

When $x = 2$, $y = (3 - 2 \times 2)^7 = (-1)^7 = -1$ *[1 mark]*

Using $y - b = m(x - a)$ to find the equation of the tangent:

$y + 1 = -14(x - 2) \Rightarrow y = -14x + 27$ *[1 mark]*

[5 marks available in total — as above]

9   $\frac{dy}{dx} = -2\sin 2x + \cos\left(x + \frac{\pi}{3}\right)$ *[1 mark for each correct term]*

At $x = -\frac{\pi}{3}$, $\frac{dy}{dx} = -2\sin\left(-\frac{2\pi}{3}\right) + \cos\left(-\frac{\pi}{3} + \frac{\pi}{3}\right)$

$= -2\left(-\frac{\sqrt{3}}{2}\right) + \cos(0) = \sqrt{3} + 1$ *[1 mark]*

[3 marks available in total — as above]

10 a) When $x = 0$, $y = (0)^3 - 7(0)^2 + 9(0) + 12 = 12$

So the curve C crosses the y-axis at $(0, 12)$. *[1 mark]*

$\frac{dy}{dx} = 3x^2 - 14x + 9$ *[1 mark]*, so when $x = 0$, $\frac{dy}{dx} = 9$ *[1 mark]*

Using $y - b = m(x - a)$ to find the equation of the tangent:

$y - 12 = 9x \Rightarrow y = 9x + 12$ *[1 mark]*

[4 marks available in total — as above]

   b) Find the intersections of $y = x^3 - 7x^2 + 9x + 12$ and $y = 9x + 12$:

$x^3 - 7x^2 + 9x + 12 = 9x + 12 \Rightarrow x^3 - 7x^2 = 0$ *[1 mark]*

$\Rightarrow x^2(x - 7) = 0$ *[1 mark]* $\Rightarrow x = 0$ or $x = 7$

So the tangent intersects the curve again at $x = 7$. *[1 mark]*

Then $y = 9 \times 7 + 12 = 75$, so the coordinates are $(7, 75)$. *[1 mark]*

[4 marks available in total — as above]

11 Write f(x) as powers of x: $\frac{x^3 - 5x^2 - 4x}{x\sqrt{x}} = x^{\frac{3}{2}} - 5x^{\frac{1}{2}} - 4x^{-\frac{1}{2}}$ *[1 mark]*

Differentiate the function: $\text{f}'(x) = \frac{3}{2}x^{\frac{1}{2}} - \frac{5}{2}x^{-\frac{1}{2}} + 2x^{-\frac{3}{2}}$

[1 mark for any two terms correct, 1 mark for the correct derivative]

so $\text{f}'(4) = \frac{3}{2}(2) - \frac{5}{2}\left(\frac{1}{2}\right) + 2\left(\frac{1}{2^3}\right) = 2$ *[1 mark]*

Using $y - b = m(x - a)$ to find the equation of the tangent:

$y + 4 = 2(x - 4) \Rightarrow y = 2x - 12$ *[1 mark]*

[5 marks available in total — as above]

Pages 45-47 — Stationary Points

1   Differentiate f(x): $\text{f}'(x) = 3x^2 - 10x + 3$ *[1 mark]*

$\text{f}'(4) = 3 \times 4^2 - 10 \times 4 + 3 = 48 - 40 + 3 = 11$ *[1 mark]*

$\text{f}'(4) > 0$, so f(x) is increasing when $x = 4$. *[1 mark]*

[3 marks available in total — as above]

2   The function is decreasing when $\frac{dy}{dx} < 0$. *[1 mark]*

$\frac{dy}{dx} = -3 - 2x$ *[1 mark]*

$-3 - 2x < 0 \Rightarrow x > -\frac{3}{2}$

So the function is decreasing for $x > -\frac{3}{2}$. *[1 mark]*

[3 marks available in total — as above]

3   The stationary points of f(x) are the points where $\text{f}'(x) = 0$, so the graph of $\text{f}'(x)$ has roots at $x = \frac{2}{3}$ and $x = 4$.

The graph of f(x) is decreasing between $x = \frac{2}{3}$ and $x = 4$, so $\text{f}'(x)$ is below the x-axis between these points.

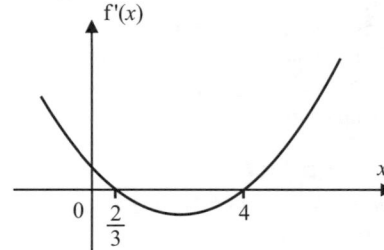

[3 marks available — 1 mark for the correct shape, 1 mark for the correct orientation, 1 mark for labelling the correct roots]

4 a) Differentiate f(x): $\text{f}'(x) = -4 - 3x^2$

[2 marks available — 1 mark for each correct term]

   b) Since $x^2 \geq 0$, $\text{f}'(x)$ has a maximum value of -4. *[1 mark]*

$\text{f}'(x) < 0$ for all x, so f(x) is strictly decreasing for all x. *[1 mark]*

[2 marks available in total — as above]

5 a) $\frac{dy}{dx} = 4x^3 - 3ax^2 - 36x + 108$ *[1 mark]*

When $x = 3$, $4(3)^3 - 3a(3)^2 - 36(3) + 108 = 0$ *[1 mark]*

$4(27) - 27a - 108 + 108 = 0$

$27a = 108 \Rightarrow a = 4$ *[1 mark]*

[3 marks available in total — as above]

   b) Using a nature table:

x	2	3	4
$\frac{dy}{dx}$	20	0	28
Slope	/	—	/

The gradient is positive on the left and the right of the stationary point, so $x = 3$ is a point of inflexion.

[2 marks available — 1 mark for a correct method, 1 mark for the correct answer]

6   $\text{f}'(x) = 3x^2 + k$ *[1 mark]*

As $x = -2$ is a stationary point, $\text{f}'(-2) = 0$ *[1 mark]*

$\text{f}'(-2) = 3(-2)^2 + k = 0 \Rightarrow k = -12$ *[1 mark]*

[3 marks available in total — as above]

7   f(x) is a strictly increasing function for all values of x if $\text{f}'(x) > 0$ for all x.

$\text{f}(x) = 3x^3 + 9x^2 + 25x \Rightarrow \text{f}'(x) = 9x^2 + 18x + 25$ *[1 mark]*

E.g. complete the square to show that $\text{f}'(x) > 0$:

$\text{f}'(x) = 9x^2 + 18x + 25 = 9(x^2 + 2x) + 25$

$= 9(x + 1)^2 - 9 + 25$

$= 9(x + 1)^2 + 16$

[1 mark for a suitable method]

$(x + 1)^2 \geq 0$, so $\text{f}'(x)$ has a minimum value of 16. *[1 mark]*

So $\text{f}'(x) > 0$ for all x, which means that f(x) is a strictly increasing function for all values of x. *[1 mark]*

[4 marks available in total — as above]

Answers

8 a) f'(x) = 8x³ + 64 *[1 mark]*

$\text{f}'(x) = 8x^3 + 64$ *[1 mark]*

At the stationary point, f'(x) = 0 $\Rightarrow$ 8x³ + 64 = 0 *[1 mark]*

$\Rightarrow x^3 = -\frac{64}{8} \Rightarrow x = \sqrt[3]{-8} = -2$ *[1 mark]*

When x = –2, f(x) = 2(–2)⁴ + 64(–2) = –96

So the stationary point is at (–2, –96) *[1 mark]*

[4 marks available in total — as above]

b) E.g. The graph of y = f(x) is a positive quartic graph with only one stationary point, so f(x) is increasing to the right of the stationary point, when x > –2, and f(x) is decreasing to the left of the stationary point, when x < –2.

[2 marks available — 1 mark for a correct method, 1 mark for the correct answers]

c) From part b), the stationary point at (–2, –96) is a minimum point. Find where the curve crosses the x-axis:

When f(x) = 0, 2x⁴ + 64x = 0 $\Rightarrow$ 2x(x³ + 32) = 0

$\Rightarrow$ x = 0 or x = $\sqrt[3]{-32}$ = –3.175 (3 d.p.)

So the curve crosses the x-axis at x = 0 and x = –3.175:

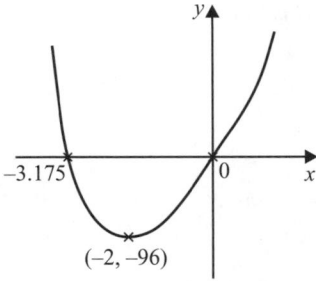

[3 marks available — 1 mark for a curve with the correct shape, 1 mark for the correct minimum point, 1 mark for the correct intercepts]

Pages 48-49 — Using Differentiation

1 $\frac{dh}{dt} = 30 - 10t$ *[1 mark]*

When the height is at its maximum, $\frac{dh}{dt} = 0$ *[1 mark]*

So 30 – 10t = 0 $\Rightarrow$ t = $\frac{30}{10}$ = 3 *[1 mark]*

Check this gives a maximum value:

t	2	3	4
$\frac{dh}{dt}$	10	0	–10
Slope	/	—	\

So the stationary point at t = 3 is a maximum. *[1 mark]*

At t = 3, h(t) = 30 × 3 – 5 × 3² = 45 m *[1 mark]*

[5 marks available in total — as above]

2 a) $\frac{dy}{dx} = 6x - 8 \times \frac{3}{2}x^{\frac{1}{2}} = 6x - 12x^{\frac{1}{2}}$ *[1 mark]*

At the stationary points, $\frac{dy}{dx} = 0$ *[1 mark]*

$\Rightarrow 6x - 12x^{\frac{1}{2}} = 0 \Rightarrow 6x^{\frac{1}{2}}(x^{\frac{1}{2}} - 2) = 0$ *[1 mark]*

So the stationary points are at x = 0 and $x^{\frac{1}{2}} = 2 \Rightarrow x = 4$ *[1 mark]*

[4 marks available in total — as above]

b) Evaluate y at the end points of the interval:

When x = 0, y = 3 × 0² – 8 × 0^{3/2} = 0

When x = 9, y = 3 × 9² – 8 × 9^{3/2} = 3 × 81 – 8 × 27 = 27

At the stationary point x = 4, y = 3 × 4² – 8 × 4^{3/2} = 48 – 64 = –16

So the greatest value on the interval is 27 and the least value on the interval is –16.

[3 marks available — 1 mark for evaluating y at both end points, 1 mark for evaluating y at the stationary point, 1 mark for stating the correct greatest and least values]

3 a) Surface area = 2 × area of circular face + area of curved face

= 2πr² + 2πrh

$\Rightarrow 2\pi r^2 + 2\pi rh = 100\pi$ *[1 mark]* $\Rightarrow h = \frac{50 - r^2}{r}$ *[1 mark]*

So volume $V = \pi r^2 h = \pi r^2 \times \frac{50 - r^2}{r} = 50\pi r - \pi r^3$ *[1 mark]*

[3 marks available in total — as above]

b) Differentiate the expression for V: $\frac{dV}{dr} = 50\pi - 3\pi r^2$

[1 mark for differentiating, 1 mark for the correct derivative]

V is a maximum when $\frac{dV}{dr} = 0$, so 50π – 3πr² = 0 *[1 mark]*

$r^2 = \frac{50\pi}{3\pi} \Rightarrow r = \sqrt{\frac{50}{3}} = 4.082...$ *[1 mark]*

Ignore the negative root here as you know r < 0.

To check that this value of r gives a maximum, use a nature table:

r	4	4.082...	5
$\frac{dV}{dr}$	2π	0	–25π
Slope	/	—	\

[1 mark]

So V is maximised when r = 4.08 cm (3 s.f.) *[1 mark]*

[6 marks available in total — as above]

You could also find the second derivative at r = 4.082... and show that it's less than 0, instead of using a nature table.

4 a) Volume of the container = length × width × height = x²y *[1 mark]*

Volume = 40 000 cm³, so: x²y = 40 000 $\Rightarrow y = \frac{40\,000}{x^2}$ *[1 mark]*

Surface area = sum of areas of all 5 sides

= x² + x² + xy + xy + xy = 2x² + 3xy

$= 2x^2 + 3x\left(\frac{40\,000}{x^2}\right)$

$= 2x^2 + \frac{120\,000}{x}$ *[1 mark]*

[3 marks available in total — as above]

b) Differentiate the expression for A:

$\frac{dA}{dx} = 4x - \frac{120\,000}{x^2}$ *[1 mark for differentiating, 1 mark for the correct derivative]*

Find the value of x where $\frac{dA}{dx} = 0$: $4x - \frac{120\,000}{x^2} = 0$ *[1 mark]*

$\Rightarrow x^3 = 30\,000 \Rightarrow x = 31.07...$ *[1 mark]*

To check if it's a minimum, use a nature table:

x	31	31.07...	32
$\frac{dA}{dx}$	–0.8699...	0	10.8125
Slope	\	—	/

[1 mark]

The slope is negative to the left of x = 31.07... and positive to the right, so the minimum is at x = 31.1 cm (3 s.f.). *[1 mark]*

[6 marks available in total — as above]

c) Put the value of x found in part b) into the formula for the area given in part a):

$A = 2 \times (31.07...)^2 + \frac{120\,000}{31.07...}$ *[1 mark]*

= 5792.936... = 5790 cm² (3 s.f.) *[1 mark]*

[2 marks available in total — as above]

5 The time taken per cookie, t, is given by the time taken to make n cookies divided by the number of cookies, n.

$t = \frac{\frac{1}{5}n^2 + 50\sqrt{n}}{n} = \frac{1}{5}n + 50n^{-\frac{1}{2}}$ *[1 mark]*

$\frac{dt}{dn} = \frac{1}{5} - 25n^{-\frac{3}{2}}$ *[1 mark for each correct term]*

At the minimum value of t, $\frac{dt}{dn} = 0 \Rightarrow \frac{1}{5} - 25n^{-\frac{3}{2}} = 0$ *[1 mark]*

$\Rightarrow 25n^{-\frac{3}{2}} = \frac{1}{5} \Rightarrow n^{-\frac{3}{2}} = \frac{1}{125} \Rightarrow n = \sqrt[3]{125^2} = 25$ *[1 mark]*

To check if it's a minimum, use a nature table:

n	24	25	26
$\frac{dt}{dn}$	–0.0126...	0	0.0114...
Slope	\	—	/

[1 mark]

The slope is negative to the left of n = 25 and positive to the right, so the time taken per cookie is at a minimum when Paul makes 25 cookies. *[1 mark]*

[7 marks available in total — as above]

Answers

Pages 50-52 — Integration

1 $\int (4x^3 + 6x + 3)\,dx = \dfrac{4x^4}{4} + \dfrac{6x^2}{2} + 3x + C = x^4 + 3x^2 + 3x + C$

[2 marks available — 1 mark for at least one term correct, 1 mark for the correct integral with + C]

2 $\int 4\sin x\,dx = 4 \times (-\cos x) + C = -4\cos x + C$

[2 marks available — 1 mark for integrating, 1 mark for the correct integral with + C]

3 Integrate the derivative to find the equation of the curve:

$y = \int (3x^2 + 6x - 4)\,dx = \dfrac{3x^3}{3} + \dfrac{6x^2}{2} - 4x + C$

$= x^3 + 3x^2 - 4x + C$

[1 mark for at least one term correct, 1 mark for the correct integral with + C]

Use the point (0, 0) to find C: $0 = 0 + 0 - 0 + C \Rightarrow C = 0$ *[1 mark]*

So the equation of the curve is $y = x^3 + 3x^2 - 4x$ *[1 mark]*

[4 marks available in total — as above]

4 Integrate the derivative to find y:

$y = \int 6\cos\left(2\left(x - \dfrac{\pi}{6}\right)\right)dx = \int 6\cos\left(2x - \dfrac{\pi}{3}\right)dx$

$= 6 \times \dfrac{1}{2}\sin\left(2x - \dfrac{\pi}{3}\right) + C$

$= 3\sin\left(2x - \dfrac{\pi}{3}\right) + C$

[1 mark for integrating, 1 mark for correct integral with + C]

Use the point $\left(\dfrac{\pi}{6}, 3\right)$ to find C:

$3 = 3\sin\left(\dfrac{2\pi}{6} - \dfrac{\pi}{3}\right) + C \Rightarrow C = 3 - 0 = 3$ *[1 mark]*

So $y = 3\sin\left(2x - \dfrac{\pi}{3}\right) + 3$ *[1 mark]*

[4 marks available in total — as above]

5 To find f(x), integrate f'(x):

$f(x) = \int \left(2x + 3\sqrt{x} + \dfrac{12}{x^2}\right)dx = \int (2x + 3x^{\frac{1}{2}} + 12x^{-2})\,dx$

$= \dfrac{2x^2}{2} + \dfrac{3x^{\frac{3}{2}}}{\left(\frac{3}{2}\right)} + \left(\dfrac{12x^{-1}}{-1}\right) + C = x^2 + 2\sqrt{x^3} - \dfrac{12}{x} + C$

Use the point (4, 17) to find C: $4^2 + 2\sqrt{4^3} - \dfrac{12}{4} + C = 17$

$\Rightarrow 16 + 16 - 3 + C = 17 \Rightarrow C = 17 - 29 = -12$

So $f(x) = x^2 + 2\sqrt{x^3} - \dfrac{12}{x} - 12$

[5 marks available — 1 mark for writing in an integrable form, 1 mark for at least one term correct, 1 mark for the correct integral with + C, 1 mark for substituting values, 1 mark for the correct answer]

6 $\int \left(\dfrac{x^2 + 3}{\sqrt{x}}\right)dx = \int (x^{\frac{3}{2}} + 3x^{-\frac{1}{2}})\,dx = \dfrac{x^{\frac{5}{2}}}{\left(\frac{5}{2}\right)} + \dfrac{3x^{\frac{1}{2}}}{\left(\frac{1}{2}\right)} + C$

$= \dfrac{2}{5}x^{\frac{5}{2}} + 6x^{\frac{1}{2}} + C = \dfrac{2}{5}\sqrt{x^5} + 6\sqrt{x} + C$

[3 marks available — 1 mark for writing in an integrable form, 1 mark for at least one term correct, 1 mark for correct integral with + C]

7 Integrate the derivative to find the equation of the curve:

$y = \int (8x + 1)^{-\frac{1}{2}}\,dx = \dfrac{1}{8} \times \dfrac{(8x + 1)^{\frac{1}{2}}}{\left(\frac{1}{2}\right)} + C$

$= \dfrac{1}{4}\sqrt{8x + 1} + C$

[1 mark for integrating, 1 mark for the correct integral with + C]

Use the point (3, 1) to find C:

$1 = \dfrac{1}{4}\sqrt{3 \times 8 + 1} + C$ *[1 mark]*

$\Rightarrow C = 1 - \dfrac{1}{4} \times 5 = 1 - \dfrac{5}{4} = -\dfrac{1}{4}$

So the equation of the curve is $y = \dfrac{1}{4}\sqrt{8x + 1} - \dfrac{1}{4}$ *[1 mark]*

[4 marks available in total — as above]

8 a) Differentiate the function to get f'(x):

$f'(x) = -\dfrac{1}{2} \times -\dfrac{4}{3} \times \left(7 - \dfrac{4}{3}x\right)^{-\frac{3}{2}} = \dfrac{2}{3}\left(7 - \dfrac{4}{3}x\right)^{-\frac{3}{2}}$

[2 marks available — 1 mark for differentiating, 1 mark for the correct derivative]

b) $-2\left(7 - \dfrac{4}{3}x\right)^{-\frac{3}{2}} = -3f'(x)$, so the integral is $-3f(x) + C$

$\Rightarrow \int -2\left(7 - \dfrac{4}{3}x\right)^{-\frac{3}{2}}dx = -3\left(7 - \dfrac{4}{3}x\right)^{-\frac{1}{2}} + C$ *[1 mark]*

9 To find f(x), integrate f'(x):

$f(x) = \int 5\sin\left(3x + \dfrac{\pi}{6}\right)dx = 5 \times \dfrac{-\cos\left(3x + \frac{\pi}{6}\right)}{3} + C$

$= -\dfrac{5}{3}\cos\left(3x + \dfrac{\pi}{6}\right) + C$

[1 mark for integrating, 1 mark for the correct integral with + C]

Use the point $\left(\dfrac{\pi}{2}, \dfrac{7}{6}\right)$ to find C:

$\dfrac{7}{6} = -\dfrac{5}{3}\cos\left(\dfrac{3\pi}{2} + \dfrac{\pi}{6}\right) + C$ *[1 mark]*

$\Rightarrow C = \dfrac{7}{6} + \dfrac{5}{3}\cos\left(\dfrac{10\pi}{6}\right)$

$\Rightarrow C = \dfrac{7}{6} + \left(\dfrac{5}{3} \times \dfrac{1}{2}\right) = \dfrac{7}{6} + \dfrac{5}{6} = 2$

So $f(x) = -\dfrac{5}{3}\cos\left(3x + \dfrac{\pi}{6}\right) + 2$ *[1 mark]*

[4 marks available in total — as above]

10 Integrate the derivative to find y:

$y = \int \dfrac{3x^2 - 5x^{\frac{1}{2}}}{x^4}\,dx = \int (3x^{-2} - 5x^{-\frac{7}{2}})\,dx$

$= \dfrac{3x^{-1}}{-1} - \left(-\dfrac{2}{5}\right)(5x^{-\frac{5}{2}}) + C = -3x^{-1} + 2x^{-\frac{5}{2}} + C$

[1 mark for writing in an integrable form, 1 mark for at least one term correct, 1 mark for the correct integral with + C]

Use the point (1, –2) to find C: $-2 = -\dfrac{3}{1} + \dfrac{2}{1} + C$ *[1 mark]*

$\Rightarrow C = -2 + 3 - 2 = -1$

So $y = -3x^{-1} + 2x^{-\frac{5}{2}} - 1$ *[1 mark]*

[5 marks available in total — as above]

11 a) $f(x) = \int 2\cos(4 - 6x)\,dx = \dfrac{2\sin(4 - 6x)}{-6} + C$

$= -\dfrac{1}{3}\sin(4 - 6x) + C$

[2 marks available — 1 mark for integrating, 1 mark for correct integral with + C]

b) Write $2\cos(4 - 6x)$ as $2\cos(2(2 - 3x))$.

Then, using the identity $\cos 2\theta \equiv 1 - 2\sin^2\theta$, *[1 mark]*

$2\cos(2(2 - 3x)) \equiv 2(1 - 2\sin^2(2 - 3x))$

$\equiv 2 - 4\sin^2(2 - 3x)$ *[1 mark]*

[2 marks available in total — as above]

c) $8\sin^2(2 - 3x) - 4 = -2 \times (2 - 4\sin^2(2 - 3x))$

$= -2(2\cos(4 - 6x))$

So $\int (8\sin^2(2 - 3x) - 4)\,dx = -2\int 2\cos(4 - 6x)\,dx$ *[1 mark]*

$= -2 \times -\dfrac{1}{3}\sin(4 - 6x) + C$

$= \dfrac{2}{3}\sin(4 - 6x) + C$ *[1 mark]*

[2 marks available in total — as above]

Pages 53-56 — Definite Integrals

1 $\int_1^3 3x^2 - 4x\,dx = \left[x^3 - \dfrac{4x^2}{2}\right]_1^3 = [x^3 - 2x^2]_1^3$

$= (3^3 - 2(3^2)) - (1 - 2(1^2))$

$= (27 - 18) - (1 - 2) = 10$

[3 marks available — 1 mark for integrating correctly, 1 mark for substituting in the limits, 1 mark for the correct answer]

2 $\int_{\frac{\pi}{12}}^{\frac{\pi}{8}} \sin 2x\,dx = -\dfrac{1}{2}[\cos 2x]_{\frac{\pi}{12}}^{\frac{\pi}{8}}$

$= -\dfrac{1}{2}\left(\cos\left(\dfrac{\pi}{4}\right) - \cos\left(\dfrac{\pi}{6}\right)\right)$

$= -\dfrac{1}{2}\left(\dfrac{\sqrt{2}}{2} - \dfrac{\sqrt{2}}{2}\right) = \dfrac{\sqrt{3} - \sqrt{2}}{4}$

[3 marks available — 1 mark for integrating correctly, 1 mark for substituting in the limits, 1 mark for the correct answer]

3   $\int_0^1 (6x+1)^{-3}\,dx = \left[\dfrac{1}{6\times -2}(6x+1)^{-2}\right]_0^1$

$= -\dfrac{1}{12}\left[(6x+1)^{-2}\right]_0^1$

$= -\dfrac{1}{12}\left([(7)^{-2}]-[(1)^{-2}]\right)$

$= -\dfrac{1}{12}\left(\dfrac{1}{49}-1\right) = -\dfrac{1}{12}\left(-\dfrac{48}{49}\right) = \dfrac{4}{49}$

[3 marks available — 1 mark for correct integral, 1 mark for substituting limits, 1 mark for the correct answer]

4   The integral is equal to the area under the curve. The three regions each have the same area, but one region is below the x-axis, so its area is negative. So $\int_0^\pi 2\sin 3x\,dx = \dfrac{4}{3}-\dfrac{4}{3}+\dfrac{4}{3} = \dfrac{4}{3}$ units²

[2 marks available — 1 mark for correct interpretation of the graph, 1 mark for the correct answer]

5   Integrate the function between $-\pi$ and π:

$\int_{-\pi}^{\pi} 5\cos\left(\dfrac{x}{6}-\pi\right)dx = \left[\dfrac{5}{\left(\frac{1}{6}\right)}\sin\left(\dfrac{x}{6}-\pi\right)\right]_{-\pi}^{\pi} = 30\left[\sin\left(\dfrac{x}{6}-\pi\right)\right]_{-\pi}^{\pi}$

$= 30\left(\sin\left(-\dfrac{5\pi}{6}\right)-\sin\left(-\dfrac{7\pi}{6}\right)\right)$

$= 30\left(-\dfrac{1}{2}-\dfrac{1}{2}\right) = -30$

[4 marks available — 1 mark for integrating, 1 mark for the correct integral, 1 mark for substituting limits, 1 mark for the correct answer]

6   To find the shaded area, integrate the function between −1 and 1 and then subtract the integral of the function between 1 and 3 (as this value will be negative).

$\int_{-1}^{1}(x^3-3x^2-6x+8)\,dx = \left[\dfrac{x^4}{4}-\dfrac{3x^3}{3}-\dfrac{6x^2}{2}+8x\right]_{-1}^{1}$

$= \left[\dfrac{x^4}{4}-x^3-3x^2+8x\right]_{-1}^{1}$

$= \left(\dfrac{(1)^4}{4}-(1)^3-3(1)+8(1)\right)$

$\quad -\left(\dfrac{(-1)^4}{4}-(-1)^3-3(-1)^2+8(-1)\right)$

$= \left(\dfrac{1}{4}-1-3+8\right)-\left(\dfrac{1}{4}+1-3-8\right)$

$= 4\dfrac{1}{4}-\left(-9\dfrac{3}{4}\right) = 14$

So the area between −1 and 1 is 14 units².

$\int_{1}^{3}(x^3-3x^2-6x+8)\,dx = \left[\dfrac{x^4}{4}-x^3-3x^2+8x\right]_{1}^{3}$

$= \left(\dfrac{(3)^4}{4}-(3)^3-3(3)^2+8(3)\right)-\left(4\dfrac{1}{4}\right)$

$= \left(\dfrac{81}{4}-27-27+24\right)-4\dfrac{1}{4}$

$= -9\dfrac{3}{4}-4\dfrac{1}{4} = -14$

So the area between 1 and 3 is −14 units².
Therefore, the total area is $14-(-14) = 28$ units².

[6 marks available — 1 mark for dealing with the area below the x-axes correctly, 1 mark for integrating with the correct limits, 1 mark for the correct integral, 1 mark for substituting the limits correctly, 1 mark for at least one area correct, 1 mark for the correct final answer]
If you'd just integrated between −1 and 3, you'd have ended up with an answer of 0, as the areas cancel each other out. You have to split the area into two parts and find each one separately.

7   Evaluate the integral, treating k as a constant:

$\int_{\sqrt{2}}^{2}(8x^3-2kx)\,dx = \left[\dfrac{8x^4}{4}-\dfrac{2kx^2}{2}\right]_{\sqrt{2}}^{2} = \left[2x^4-kx^2\right]_{\sqrt{2}}^{2}$

$= (2(2)^4-k(2)^2)-(2(\sqrt{2})^4-k(\sqrt{2})^2)$

$= (32-4k)-(8-2k) = 24-2k$

You know that the value of this integral is $2k^2$:

$24-2k = 2k^2 \Rightarrow 2k^2+2k-24 = 0$

$\Rightarrow k^2+k-12 = 0$

$\Rightarrow (k+4)(k-3) = 0$

$\Rightarrow k = -4$ or $k = 3$

[6 marks available — 1 mark for integrating, 1 mark for the correct integration, 1 mark for substituting in limits, 1 mark for setting equal to 2k², 1 mark for factorising quadratic, 1 mark for both values of k]

8   Area $= \int_0^1\left(\left(\sqrt{x}-\dfrac{1}{2}x^2+1\right)-\left(2x-\dfrac{1}{2}\right)\right)dx$

$= \int_0^1\left(x^{\frac{1}{2}}-\dfrac{1}{2}x^2-2x+\dfrac{3}{2}\right)dx$

$= \left[\dfrac{2}{3}x^{\frac{3}{2}}-\dfrac{1}{6}x^3-x^2+\dfrac{3}{2}x\right]_0^1$

$= \left(\dfrac{2}{3}-\dfrac{1}{6}-1+\dfrac{3}{2}\right)-0 = 1\,\text{unit}^2$

[5 marks available — 1 mark for a suitable method to find the area between the line and the curve, 1 mark for the correct limits, 1 mark for integrating, 1 mark for substituting the limits correctly, 1 mark for the correct answer]
You could also have done this one by working out the area under each curve separately and subtracting them.

9   Find the x-coordinate of the intersection of $c(x)$ and $d(x)$

$c(x) = d(x) \Rightarrow x-2 = -x+4 \Rightarrow 2x = 6 \Rightarrow x = 3$

The shape is symmetrical, so the area between $a(x)$ and $d(x)$ is equal to the area between $b(x)$ and $c(x)$.
Integrate to find the area between $a(x)$ and $d(x)$:

$\int_0^3\left(\left(\sin\left(\dfrac{\pi x}{3}\right)+4\right)-(-x+4)\right)dx = \int_0^3\left(\sin\left(\dfrac{\pi x}{3}\right)+x\right)dx$

$= \left[-\dfrac{3}{\pi}\cos\left(\dfrac{\pi x}{3}\right)+\dfrac{x^2}{2}\right]_0^3$

$= \left(\dfrac{3}{\pi}+\dfrac{9}{2}\right)-\left(-\dfrac{3}{\pi}\right)$

$= \dfrac{6}{\pi}+\dfrac{9}{2} = 6.409...$

So the area of the heart is: $2\times 6.409... = 12.8$ units² (3 s.f.)

[7 marks available — 1 mark for finding the x-coordinate of the line of symmetry, 1 mark for a suitable method to find the area between a curve and line, 1 mark for the correct limits, 1 mark for integrating, 1 mark for substituting the limits correctly, 1 mark for at least one area correct, 1 mark for the correct total area]

10 a)   Find the points where $y = 3x^2-6x+5$ and $y = 14$ intersect:

$3x^2-6x+5 = 14$ *[1 mark]* $\Rightarrow 3x^2-6x-9 = 0$

$\Rightarrow x^2-2x-3 = 0$

$\Rightarrow (x-3)(x+1) = 0$ *[1 mark]*

So the limits are $x = -1$ and $x = 3$.

Area $= \int_{-1}^{3}(14-(3x^2-6x+5))\,dx$ *[1 mark]*

$= \int_{-1}^{3}9+6x-3x^2\,dx$

$= [9x+3x^2-x^3]_{-1}^{3}$ *[1 mark]*

$= [(27+27-27)-(-9+3+1)]$ *[1 mark]*

$= 27-(-5) = 32$ units² *[1 mark]*

[6 marks available in total — as above]

b)   The area to the left of the x-axis is bounded by the curves $y = 3x^2-6x+5$ and $y = 14$, and the line $x = 0$.

Area $= \int_{-1}^{0}9+6x-3x^2\,dx$

$= [9x+3x^2-x^3]_{-1}^{0}$

$= [0-(-9+3+1)]$

$= 0-(-5) = 5$

The rest of the area $(32-5 = 27)$ is to the right of the x-axis, so the ratio is $5:27$.

[4 marks available — 1 mark for correct limits for the region on either side of the y-axis, 1 mark for integrating correctly, 1 mark for substituting in the limits and evaluating the area, 1 mark for the correct ratio]

11   The shaded region is the area between the curves between $x = 1$ and $x = 3$:

$\int_1^3(10-x^2-9x^{-2})\,dx = \left[10x-\dfrac{x^3}{3}+9x^{-1}\right]_1^3$

$= \left(10(3)-\dfrac{(3)^3}{3}+9(3)^{-1}\right)-\left(10-\dfrac{1}{3}+9\right)$

$= (30-9+3)-\left(\dfrac{56}{3}\right)$

$= 24-\dfrac{56}{3} = \dfrac{16}{3}$ units²

[6 marks available — 1 mark for writing in an integrable form, 1 mark for a suitable method for finding the area between the curves, 1 mark for the correct limits, 1 mark for integrating, 1 mark for substituting the limits correctly, 1 mark for the correct answer]

Answers

Pages 57-58 — Rates of Change

1. To find the rate of change, differentiate the function:
$f(x) = (3x + 1)^{-2}$
So $f'(x) = -2 \times (3x + 1)^{-3} \times 3 = -\dfrac{6}{(3x + 1)^3}$
[1 mark for writing in a differentiable form, 1 mark for the correct derivative]
$f'(3) = -\dfrac{6}{(3 \times (-1) + 1)^3} = -\dfrac{6}{(-2)^3} = \dfrac{3}{4}$ *[1 mark]*
[3 marks available in total — as above]

2. a) Velocity = rate of change of displacement = $\dfrac{dx}{dt}$, so to find the formula for velocity, differentiate the equation for x:
$v = \dfrac{dx}{dt} = 6t^2 - 8t$
[2 marks available — 1 mark for differentiating, 1 mark for the correct derivative]

   b) Find t by solving the equation $30 = 6t^2 - 8t$:
$6t^2 - 8t - 30 = 0$ *[1 mark]* $\Rightarrow 3t^2 - 4t - 15 = 0$
$\Rightarrow (3t + 5)(t - 3) = 0$
$\Rightarrow t = -\dfrac{5}{3}$ or $t = 3$
Since $t \geq 0$, $t = 3$ s *[1 mark]*
Then substitute $t = 3$ into the equation for x:
$x = 2(3)^3 - 4(3)^2 = 54 - 36 = 18$ m *[1 mark]*
[3 marks available in total — as above]

   c) Acceleration = rate of change of velocity = $\dfrac{dv}{dt}$, so find the formula for acceleration by differentiating the equation for v:
$\dfrac{dv}{dt} = 12t - 8$. So after 4 seconds, $\dfrac{dv}{dt} = 12 \times 4 - 8$
$= 48 - 8 = 40$ m/s^2
[3 marks available — 1 mark for differentiating, 1 mark for the correct derivative, 1 mark for the correct answer]

3. To find the rate of change, differentiate the function:
$P'(x) = 4 \times 4 (\cos x)^3 \times -\sin x = -16 (\cos x)^3 \sin x$
So $P'\left(\dfrac{\pi}{4}\right) = -16 \cos^3\left(\dfrac{\pi}{4}\right) \sin\left(\dfrac{\pi}{4}\right) = -16 \left(\dfrac{\sqrt{2}}{2}\right)^3 \left(\dfrac{\sqrt{2}}{2}\right)$
$= -16 \dfrac{2\sqrt{2} \times \sqrt{2}}{8 \times 2} = -16 \dfrac{4}{16} = -4$
[3 marks available — 1 mark for differentiating, 1 mark for the correct derivative, 1 mark for the correct answer]

4. a) The volume of the siren is increasing if the rate of change is positive. *[1 mark]*
$\dfrac{dV}{dt}\left(\dfrac{\pi}{8}\right) = 64 \sin\left(4\dfrac{\pi}{8} + \dfrac{\pi}{3}\right) = 64 \sin\left(\dfrac{5\pi}{6}\right) = 32 > 0$ *[1 mark]*
So the volume of the siren is increasing, as required.
[2 marks available in total — as above]

   b) You can find a formula for the volume of the siren by integrating the formula for rate of change.
$V = \int 64 \sin\left(4t + \dfrac{\pi}{3}\right) dt$
$= -64 \left(\dfrac{1}{4}\right) \cos\left(4t + \dfrac{\pi}{3}\right) + C$
$= -16 \cos\left(4t + \dfrac{\pi}{3}\right) + C$
Since $V = 120$ when $t = \dfrac{\pi}{6}$:
$120 = -16 \cos\left(\dfrac{4\pi}{6} + \dfrac{\pi}{3}\right) + C \Rightarrow 120 = -16 \cos \pi + C$
$\Rightarrow 120 = 16 + C$
$\Rightarrow C = 104$
So $V = -16 \cos\left(4t + \dfrac{\pi}{3}\right) + 104$
[4 marks available — 1 mark for integrating, 1 mark for the correct integral with + C, 1 mark for substituting values, 1 mark for the correct answer]

   c) At $t = \dfrac{\pi}{2}$, $V = -16 \cos\left(4\dfrac{\pi}{2} + \dfrac{\pi}{3}\right) + 104$ *[1 mark]*
$= -16 \cos\left(\dfrac{7\pi}{3}\right) + 104$
$= (-16)\left(\dfrac{1}{2}\right) + 104 = 96$ dB *[1 mark]*
[2 marks available in total — as above]

5. To find H in terms of t, integrate the function $\dfrac{dH}{dt}$:
$H = \int \left(\dfrac{3}{4} t^{\frac{1}{2}} + k\right) dt = \dfrac{3}{4} \times \dfrac{2}{3} t^{\frac{3}{2}} + kt + C$
$= \dfrac{1}{2}\sqrt{t^3} + kt + C$
[1 mark for integrating, 1 mark for the correct integral with + C]
When $t = 0$, $H = 0$, so $0 = 0 + 0 + C \Rightarrow C = 0$ *[1 mark]*
When $t = 4$, $H = 10$, so $10 = \dfrac{1}{2}\sqrt{4^3} + 4k = \dfrac{1}{2}\sqrt{64} + 4k$ *[1 mark]*
$\Rightarrow 10 = \dfrac{1}{2} \times 8 + 4k \Rightarrow 6 = 4k \Rightarrow k = \dfrac{3}{2}$ *[1 mark]*
So $H = \dfrac{1}{2} t^{\frac{3}{2}} + \dfrac{3}{2} t$ *[1 mark]*
[6 marks available in total — as above]

Practice Paper 1

1. a) $32 = 2^5$, so $\log_2 32 = 5$ *[1 mark]*

   b) $16 = 4^2$, so $\log_4 16 + \log_{16} 4 = 2 + 0.5 = 2.5$
   [2 marks available — 2 marks for the correct answer, otherwise 1 mark for finding the value of one of the logs correctly]

2. The radius of the circle is:
$\sqrt{(-3 - 1)^2 + (6 - (-3))^2} = \sqrt{(-4)^2 + 9^2} = \sqrt{16 + 81} = \sqrt{97}$ *[1 mark]*
So the equation of the circle is $(x - 1)^2 + (y + 3)^2 = 97$ *[1 mark]*
[2 marks available in total — as above]
You could give your answer in a different form, e.g. $x^2 - 2x + y^2 + 6y - 87 = 0$.

3. $4x^2 - 8x + 9 = 4(x^2 - 2x) + 9$ *[1 mark]*
$= 4((x - 1)^2 - 1) + 9$ *[1 mark]*
$= 4(x - 1)^2 - 4 + 9 = 4(x - 1)^2 + 5$ *[1 mark]*
[3 marks available in total — as above]

4. $\dfrac{\cos 2x}{\cos x} + \sin x \tan x \equiv \dfrac{\cos^2 x - \sin^2 x}{\cos x} + \sin x \dfrac{\sin x}{\cos x}$ *[1 mark]*
$\equiv \dfrac{\cos^2 x}{\cos x} - \dfrac{\sin^2 x}{\cos x} + \dfrac{\sin^2 x}{\cos x} \equiv \cos x$ *[1 mark]*
[2 marks available in total — as above]
There are other valid ways that you could have proved this.

5. a) $g(f(x)) = 8(f(x))^3 + 3 = 8(2 \cos(3x - 1))^3 + 3$ *[1 mark]*
$= 64 \cos^3(3x - 1) + 3$ *[1 mark]*
[2 marks available in total — as above]

   b) Let $y = g(x) \Rightarrow y = 8x^3 + 3$ *[1 mark]*
$\Rightarrow 8x^3 = y - 3 \Rightarrow x^3 = \dfrac{y - 3}{8}$ *[1 mark]*
$x = \sqrt[3]{\dfrac{y - 3}{8}}$ or $\dfrac{1}{2}\sqrt[3]{y - 3}$
$\Rightarrow g^{-1}(x) = \sqrt[3]{\dfrac{x - 3}{8}}$ or $\dfrac{1}{2}\sqrt[3]{x - 3}$ *[1 mark]*
[3 marks available in total — as above]

6. a) $\sin t$ has period 2π, so $\sin 2t$ has period $\dfrac{2\pi}{2} = \pi$
So $s(t)$ has period π seconds. *[1 mark]*
If no angle units are given in the question, you should assume that it's measured in radians.

   b) The maximum and minimum values of $\sin 2t$ are 1 and -1, so the maximum value of $s(t)$ is $4(1) - 1 = 4 - 1 = 3$ *[1 mark]*
and the minimum value of $s(t)$ is $4(-1) - 1 = -4 - 1 = -5$ *[1 mark]*
[2 marks available in total — as above]

7. If $x^2 - (m + 2)x + (6 - m) = 0$ has equal roots, then the discriminant is equal to 0. *[1 mark]*
$b^2 - 4ac = (m + 2)^2 - 4 \times 1 \times (6 - m)$
$= m^2 + 4m + 4 - 24 + 4m = m^2 + 8m - 20$ *[1 mark]*
$m^2 + 8m - 20 = 0 \Rightarrow (m + 10)(m - 2) = 0$ *[1 mark]*
So h(x) has equal roots when $m = -10$ or $m = 2$. *[1 mark]*
[4 marks available in total — as above]

8 a) $y = x^4 - 6x^2 - 8x + 5 \Rightarrow \dfrac{dy}{dx} = 4x^3 - 12x - 8$

[2 marks available — 2 marks for all three correct terms only, otherwise 1 mark for any two correct terms]

b) $\dfrac{dy}{dx} = 0$ when $x = -1$, so divide $4x^3 - 12x - 8$ by $(x + 1)$:

−1	4	0	−12	−8
		−4	4	8
	4	−4	−8	0

So $4x^3 - 12x - 8 = (x + 1)(4x^2 - 4x - 8) = 4(x + 1)(x^2 - x - 2)$
$\qquad\qquad\qquad\qquad = 4(x + 1)(x + 1)(x - 2)$

So the graph has another stationary point at $x = 2$.

[3 marks available — 1 mark for a suitable method to factorise $4x^3 - 12x - 8$, 1 mark for a correct factorisation, 1 mark for the correct answer]

c) E.g. Check the value of $\dfrac{dy}{dx}$ either side of $x = -1$:

At $x = 0$, $\dfrac{dy}{dx} = 4(0)^3 - 12(0) - 8 = -8$

At $x = -2$, $\dfrac{dy}{dx} = 4(-2)^3 - 12(-2) - 8 = -32 + 24 - 8 = -16$

These are both negative, so the stationary point is a (falling) point of inflexion.

[2 marks available — 1 mark for a correct method or justification, 1 mark for the correct answer]

You could also have worked out that it was a point of inflexion based on the fact that $\dfrac{dy}{dx}$ has a double root at $x = -1$, or based on the shape of the graph — just as long as you properly justify your answer.

9 $\sin(x + y) \equiv \sin x \cos y + \cos x \sin y$ *[1 mark]*

From the diagram, $\sin x = \dfrac{3}{5}$, $\cos x = \dfrac{4}{5}$,

$\sin y = \dfrac{1}{\sqrt{10}}$ and $\cos y = \dfrac{3}{\sqrt{10}}$ *[1 mark for all four]*

So $\sin(x + y) = \left(\dfrac{3}{5} \times \dfrac{3}{\sqrt{10}}\right) + \left(\dfrac{4}{5} \times \dfrac{1}{\sqrt{10}}\right)$

$\qquad = \dfrac{9}{5\sqrt{10}} + \dfrac{4}{5\sqrt{10}} = \dfrac{13}{5\sqrt{10}} = \dfrac{13\sqrt{10}}{50}$ *[1 mark]*

[3 marks available in total — as above]

10 a) L_1 is the altitude from Q, so L_1 is perpendicular to PR.

Gradient of PR $= \dfrac{(-4) - 2}{8 - (-4)} = \dfrac{-6}{12} = -\dfrac{1}{2}$ *[1 mark]*

So gradient of $L_1 = -1 \div -\dfrac{1}{2} = 2$ *[1 mark]*

Then the equation of L_1 is:

$(y - 5) = 2(x - 5) \Rightarrow y = 2x - 10 + 5 \Rightarrow y = 2x - 5$ *[1 mark]*

[3 marks available in total — as above]

You could also give your answer in a different form, e.g. $2x - y - 5 = 0$. As long as you've simplified so there's only one constant term, you'll get the marks.

b) $\tan a = \dfrac{1}{3} \Rightarrow$ gradient of perpendicular bisector of PS $= \dfrac{1}{3}$,

and since it passes through the origin, the perpendicular bisector of PS has equation $y = \dfrac{1}{3}x$ *[1 mark]*

So gradient of PS $= -1 \div \dfrac{1}{3} = -3$ *[1 mark]*

Then the equation of PS is:

$(y - 2) = -3(x - (-4)) \Rightarrow y = -3x - 12 + 2$
$\qquad\qquad\qquad\qquad \Rightarrow y = -3x - 10$ *[1 mark]*

[3 marks available in total — as above]

c) L_1 and the perpendicular bisector of PS intersect where $y = 2x - 5$ crosses $y = \dfrac{1}{3}x$:

$2x - 5 = \dfrac{1}{3}x \Rightarrow 6x - 15 = x \Rightarrow 5x = 15 \Rightarrow x = 3$ *[1 mark]*

$y = \dfrac{1}{3}x = \dfrac{1}{3} \times 3 = 1$

So the lines cross at $(3, 1)$ *[1 mark]*

[2 marks available in total — as above]

11 $\displaystyle\int_0^{\frac{\pi}{6}} 2\sin\left(3x - \dfrac{\pi}{2}\right) dx = \left[-\dfrac{2}{3}\cos\left(3x - \dfrac{\pi}{2}\right)\right]_0^{\frac{\pi}{6}}$

[1 mark for $-\cos\left(3x - \dfrac{\pi}{2}\right)$ and 1 mark for $\dfrac{2}{3}$]

$= -\dfrac{2}{3}\left[\cos\left(3x - \dfrac{\pi}{2}\right)\right]_0^{\frac{\pi}{6}}$

$= -\dfrac{2}{3}\left(\left(\cos\left(\dfrac{\pi}{2} - \dfrac{\pi}{2}\right)\right) - \left(\cos\left(0 - \dfrac{\pi}{2}\right)\right)\right)$ *[1 mark]*

$= -\dfrac{2}{3}\left(\cos 0 - \cos\left(-\dfrac{\pi}{2}\right)\right) = -\dfrac{2}{3}(1 - 0) = -\dfrac{2}{3}$ *[1 mark]*

[4 marks available in total — as above]

12 $\overrightarrow{AB} = (-2 - (-7))\mathbf{i} + (2 - 4)\mathbf{j} + (a - 3)\mathbf{k} = 5\mathbf{i} - 2\mathbf{j} + (a - 3)\mathbf{k}$ *[1 mark]*
$\overrightarrow{BC} = (0 - (-2))\mathbf{i} + (1 - 2)\mathbf{j} + ((-1) - a)\mathbf{k} = 2\mathbf{i} - \mathbf{j} - (1 + a)\mathbf{k}$ *[1 mark]*

If $\overrightarrow{AB}$ and $\overrightarrow{BC}$ are perpendicular, then $\overrightarrow{AB}.\overrightarrow{BC} = 0$. *[1 mark]*

$\overrightarrow{AB}.\overrightarrow{BC} = (5 \times 2) + ((-2) \times (-1)) - (a - 3)(1 + a) = 0$ *[1 mark]*
$\Rightarrow 10 + 2 - (a^2 - 2a - 3) = 0 \Rightarrow -a^2 + 2a + 15 = 0$
$\Rightarrow a^2 - 2a - 15 = 0 \Rightarrow (a + 3)(a - 5)$ *[1 mark]*

So $\overrightarrow{AB}$ and $\overrightarrow{BC}$ are perpendicular when $a = -3$ or $a = 5$. *[1 mark]*
[6 marks available in total — as above]
You could also write these using column vectors, rather than i, j and k.

13 $\dfrac{1}{3}f(2x) - 1$ is a horizontal stretch by scale factor $\dfrac{1}{2}$, followed by a vertical stretch by scale factor $\dfrac{1}{3}$, then a translation by 1 unit down.

The point $(0, -3)$ is unaffected by the horizontal stretch, is moved to $(0, -1)$ by the vertical stretch and is then shifted down to $(0, -2)$.

The point $(6, 9)$ is moved to $(3, 9)$ by the horizontal stretch, then moved to $(3, 3)$ by the vertical stretch, and finally shifted down to $(3, 2)$.

So the graph looks like this:

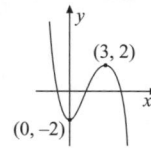

[3 marks available — 1 mark for the first turning point at $(0, -2)$, 1 mark for the second turning point having x-coordinate 3, 1 mark for the second turning point having y-coordinate 2]

14 When $y = ax + 1$ and $(x - 3)^2 + (y - 2)^2 = 5$ intersect:
$(x - 3)^2 + ((ax + 1) - 2)^2 = 5$ *[1 mark]*
$\Rightarrow (x - 3)^2 + (ax - 1)^2 = 5$
$\Rightarrow x^2 - 6x + 9 + a^2x^2 - 2ax + 1 - 5 = 0$
$\Rightarrow (1 + a^2)x^2 + (-6 - 2a)x + 5 = 0$ *[1 mark]*

If the line is a tangent to the circle, then this quadratic will only have one solution, meaning the discriminant will be equal to zero. *[1 mark]*

$b^2 - 4ac = 0 \Rightarrow (-6 - 2a)^2 - 4(1 + a^2)(5) = 0$ *[1 mark]*
$\qquad\qquad \Rightarrow 36 + 24a + 4a^2 - 20 - 20a^2 = 0$
$\qquad\qquad \Rightarrow -16a^2 + 24a + 16 = 0$
$\qquad\qquad \Rightarrow 2a^2 - 3a - 2 = 0$
$\qquad\qquad \Rightarrow (2a + 1)(a - 2) = 0$ *[1 mark]*

So the line is a tangent to the circle when $a = -\dfrac{1}{2}$ or $a = 2$ *[1 mark]*
[6 marks available in total — as above]

15 a) The volume of the plastic part is:
$V = (2x \times 2x \times y) - (x \times x \times y) = 4x^2y - x^2y = 3x^2y$ cm³
The volume is fixed at 36 cm³, so
$3x^2y = 36 \Rightarrow y = \dfrac{36}{3x^2} = \dfrac{12}{x^2}$ *[1 mark]*

The surface area of the plastic part is:
$A = 2(\text{L-shaped face}) + 2(\text{large side face}) + 4(\text{small side face})$
$= 2((2x \times 2x) - (x \times x)) + 2(2x \times y) + 4(x \times y)$ *[1 mark]*
$= 2(4x^2 - x^2) + 2(2xy) + 4(xy)$
$= 2(3x^2) + 4xy + 4xy = 6x^2 + 8xy$ *[1 mark]*

Substitute in $y = \dfrac{12}{x^2}$:
$A = 6x^2 + \left(8x \times \dfrac{12}{x^2}\right) = 6x^2 + \dfrac{96x}{x^2} = 6x^2 + \dfrac{96}{x}$ *[1 mark]*
[4 marks available in total — as above]

b) $A = 6x^2 + 96x^{-1} \Rightarrow \dfrac{dA}{dx} = 12x - 96x^{-2} = 12x - \dfrac{96}{x^2}$
[1 mark for each correct term]

When A is at a minimum, $\dfrac{dA}{dx} = 0$ *[1 mark]*
$\Rightarrow 12x - \dfrac{96}{x^2} = 0 \Rightarrow 12x = \dfrac{96}{x^2}$
$\Rightarrow 12x^3 = 96 \Rightarrow x^3 = 8 \Rightarrow x = 2$ *[1 mark]*
Verify that this is a minimum:
When $x = 1$, $\dfrac{dA}{dx} = 12(1) - \dfrac{96}{1^2} = 12 - 96 = -84$
When $x = 3$, $\dfrac{dA}{dx} = 12(3) - \dfrac{96}{3^2} = 36 - \dfrac{96}{9}$
$= 36 - 10\tfrac{2}{3} = 25\tfrac{1}{3}$
[1 mark for a suitable method]
$\dfrac{dA}{dx}$ changes from negative to positive,
so $x = 2$ is a minimum point. *[1 mark]*
When $x = 2$, $A = 6(2)^2 + \dfrac{96}{2} = 6 \times 4 + 48$
$= 24 + 48 = 72$ cm² *[1 mark]*
[7 marks available in total — as above]

Practice Paper 2

1 The circle $x^2 + y^2 - 16x - 10y + 69 = 0$ has centre (8, 5) *[1 mark]*
The line from the centre to A has a gradient of $\dfrac{7-5}{4-8} = -\dfrac{1}{2}$ *[1 mark]*
so the tangent has a gradient of $-1 \div -\dfrac{1}{2} = 2$ *[1 mark]*
So the equation of the tangent is:
$(y - 7) = 2(x - 4) \Rightarrow y = 2x - 1$ *[1 mark]*
[4 marks available in total — as above]

2 If $f(x) > -6$, $x^2 - 6x + 2 > -6 \Rightarrow x^2 - 6x + 8 > 0$
$\Rightarrow (x - 2)(x - 4) > 0$ *[1 mark]*
$x^2 - 6x + 8$ has a positive coefficient of x^2,
so the graph is u-shaped. *[1 mark]*
So $x^2 - 6x + 8$ is positive for $x < 2$ or $x > 4$. *[1 mark]*
[3 marks available in total — as above]

3 a) $u_4 = (k - 1)u_3 + 3 = (k - 1)(2k + 2) + 3$ *[1 mark]*
$= 2k^2 - 2k + 2k - 2 + 3 = 2k^2 + 1$ *[1 mark]*
[2 marks available in total — as above]

b) $2k^2 + 1 = 6.12 \Rightarrow 2k^2 = 5.12 \Rightarrow k^2 = 2.56$
k is a positive constant, so $k = 1.6$ *[1 mark]*

c) The limit L satisfies $L = (k - 1)L + 3$ *[1 mark]*
$\Rightarrow L = (1.6 - 1)L + 3 \Rightarrow L = 0.6L + 3$
$\Rightarrow 0.4L = 3 \Rightarrow L = 3 \div 0.4 = 7.5$ *[1 mark]*
[2 marks available in total — as above]

4 The graph of $g(x) = af(bx)$ is the graph of $f(x)$ stretched by scale factor a in the y-direction and scale factor $\dfrac{1}{b}$ in the x-direction.
The point (3, 3) is mapped to the point (6, 9), so:
$a = 9 \div 3 = 3$ *[1 mark]* and $\dfrac{1}{b} = 6 \div 3 \Rightarrow b = \dfrac{1}{2}$ *[1 mark]*
The graph of $h(x) = f(x + c) + d$ is the graph of $f(x)$
translated c units to the left and d units up.
The point (3, 3) is mapped to the point (0, 0), so:
$c = 3$ *[1 mark]* and $d = -3$ *[1 mark]*
[4 marks available in total — as above]

5 a) When $x = 0$, $y = \dfrac{3}{4\sqrt{(5(0) + 1)^3}} = \dfrac{3}{4\sqrt{1^3}} = \dfrac{3}{4}$ *[1 mark]*
$y = \dfrac{3}{4\sqrt{(5x + 1)^3}} = \dfrac{3}{4(5x + 1)^{\frac{3}{2}}} = \dfrac{3}{4}(5x + 1)^{-\frac{3}{2}}$ *[1 mark]*
$\dfrac{dy}{dx} = \dfrac{3}{4} \times \left(-\dfrac{3}{2}\right) \times 5 \times (5x + 1)^{-\frac{5}{2}} = -\dfrac{45}{8}(5x + 1)^{-\frac{5}{2}}$ *[1 mark]*
When $x = 0$, $\dfrac{dy}{dx} = -\dfrac{45}{8}(5(0) + 1)^{-\frac{5}{2}} = -\dfrac{45}{8}(1)^{-\frac{5}{2}} = -\dfrac{45}{8}$ *[1 mark]*
So the equation of the tangent is:
$y = -\dfrac{45}{8}x + \dfrac{3}{4}$ *[1 mark]*
[5 marks available in total — as above]

b) Shaded area $= \displaystyle\int_0^3 \dfrac{3}{4\sqrt{(5x + 1)^3}}\, dx$ *[1 mark for the correct limits]*
$= \displaystyle\int_0^3 \dfrac{3}{4}(5x + 1)^{-\frac{3}{2}}\, dx$ *[1 mark]*
$= \left[\dfrac{3}{4} \times (-2) \times \dfrac{1}{5} \times (5x + 1)^{-\frac{1}{2}}\right]_0^3 = \left[\dfrac{-3}{10\sqrt{5x + 1}}\right]_0^3$ *[1 mark]*
$= \left(\dfrac{-3}{10\sqrt{5(3) + 1}}\right) - \left(\dfrac{-3}{10\sqrt{5(0) + 1}}\right)$ *[1 mark]*
$= \dfrac{-3}{40} + \dfrac{3}{10} = \dfrac{9}{40}$ units² *[1 mark]*
[5 marks available in total — as above]

6 a) $V = V_0e^{rt} \Rightarrow \ln V = \ln(V_0e^{rt}) = \ln V_0 + \ln e^{rt} = \ln V_0 + rt$ *[1 mark]*
When $t = 0$, $\ln V = 2.5 \Rightarrow \ln V_0 + r(0) = 2.5$ *[1 mark]*
$\Rightarrow \ln V_0 = 2.5 \Rightarrow V_0 = e^{2.5} = 12.182...$ *[1 mark]*
When $t = 12$, $\ln V = 4 \Rightarrow \ln V_0 + r(12) = 4$ *[1 mark]*
$\Rightarrow 2.5 + 12r = 4 \Rightarrow 12r = 1.5$
$\Rightarrow r = 1.5 \div 12 = 0.125 \left(= \dfrac{1}{8}\right)$ *[1 mark]*
[5 marks available in total — as above]
You could also work out r and ln V_0 by finding the gradient and y-intercept of the line respectively — you'd get the same answers.

b) When $V > W$, $V_0e^{rt} > 20e^{0.08t} \Rightarrow e^{2.5}e^{0.125t} > 20e^{0.08t}$ *[1 mark]*
$\Rightarrow e^{2.5} \times e^{0.125t} \div e^{0.08t} > 20 \Rightarrow e^{2.5 + 0.125t - 0.08t} > 20$ *[1 mark]*
$\Rightarrow e^{2.5 + 0.045t} > 20 \Rightarrow 2.5 + 0.045t > \ln 20$ *[1 mark]*
$\Rightarrow 0.045t > 2.995... - 2.5 \Rightarrow t > 0.495... \div 0.045 = 11.016...$

So BankCorp is first worth more than Enterprise Inc.
when $t = 11.02$ (2 d.p.) *[1 mark]*
[4 marks available in total — as above]
You could also find ln W and solve the equation ln V > ln W.

7 $y = \cos^3 x = (\cos x)^3$, so let $u = \cos x \Rightarrow y = u^3$
Then $\dfrac{dy}{du} = 3u^2$ and $\dfrac{du}{dx} = -\sin x$ *[1 mark]*
Using the chain rule, $\dfrac{dy}{dx} = \dfrac{dy}{du} \times \dfrac{du}{dx} = (3u^2) \times (-\sin x)$
$= -3\cos^2 x \sin x$ *[1 mark]*
[2 marks available in total — as above]

8 a) (i) The distance in the y-direction from R to S is $(16 - 10) = 6$
units, and from S to T is $(10 - (-5)) = 15$ units.
So S divides the line RT in the ratio $6 : 15 = 2 : 5$. *[1 mark]*

(ii) Since S divides RT in the ratio 2 : 5, S is $\dfrac{2}{7}$ of the way
from R to T. The distance in the x-direction from R to T
is $15 - (-13) = 28$ units, so S is $\dfrac{2 \times 28}{7} = 8$ units from R
in the x-direction, meaning $k = (-13) + 8 = -5$. *[1 mark]*

b) $r_S = r_R + r_T =$ length of RT $= \sqrt{28^2 + (-21)^2} = \sqrt{784 + 441}$
$= \sqrt{1225} = 35$ *[1 mark]*
S divides RT in the ratio 2 : 5 from part a),
so length of RS $= 35 \div 7 \times 2 = 10$ *[1 mark]*
C_R touches C_S internally, so the length of RS $= r_S - r_R$ *[1 mark]*
$\Rightarrow 10 = 35 - r_R \Rightarrow r_R = 25$ *[1 mark]*
So the point where C_R and C_T meet is 25 units along the line RT,
which is $\dfrac{25}{35} = \dfrac{5}{7}$ of the way along line RT. This means it's
$28 \div 7 \times 5 = 20$ units from R in the x-direction, and
$-21 \div 7 \times 5 = -15$ units from R in the y-direction. So the
coordinates of this point are $(-13 + 20, 16 - 15) = (7, 1)$ *[1 mark]*
[5 marks available in total — as above]
You could also have found r_T rather than r_R, and used this to find the distance from T to the required point. You'd get the same answer either way.

9 Using the addition formulas:

$\sin\left(x + \frac{\pi}{3}\right) \equiv \sin x \cos \frac{\pi}{3} + \cos x \sin \frac{\pi}{3} = \frac{1}{2}\sin x + \frac{\sqrt{3}}{2}\cos x$ *[1 mark]*

$\cos\left(x - \frac{\pi}{2}\right) \equiv \cos x \cos \frac{\pi}{2} + \sin x \sin \frac{\pi}{2} = \sin x$ *[1 mark]*

So $2\sin\left(x + \frac{\pi}{3}\right) = 3\cos\left(x - \frac{\pi}{2}\right) \Rightarrow \sin x + \sqrt{3}\cos x = 3\sin x$ *[1 mark]*

$\Rightarrow 2\sin x = \sqrt{3}\cos x \Rightarrow \tan x = \frac{\sqrt{3}}{2}$ *[1 mark]* $\Rightarrow x = 0.7137...$

So in the interval $0 \le x \le 2\pi$, the solutions are:
$x = 0.714$ (3 d.p.) *[1 mark]* and $x = 3.855$ (3 d.p.) *[1 mark]*
[6 marks available in total — as above]

10 To find n, integrate $\frac{dn}{dt}$ with respect to t:

$\int \frac{3000}{(t+1)^4}\,dt = \int 3000(t+1)^{-4}\,dt$ *[1 mark]*

$= 3000 \times \left(-\frac{1}{3}\right) \times (t+1)^{-3} + C$

$= -1000(t+1)^{-3} + C = -\frac{1000}{(t+1)^3} + C$ *[1 mark]*

When $t = 0$, $n = 500$, so: $-\frac{1000}{(0+1)^3} + C = 500$ *[1 mark]*

$\Rightarrow C = 500 + \frac{1000}{1^3} = 1500$

So $n = 1500 - \frac{1000}{(t+1)^3}$ *[1 mark]*
[4 marks available in total — as above]

11 a) Using the addition formula:
$p\sin(t+q) = p\sin t \cos q + p\cos t \sin q$ *[1 mark]*
So if $A(t) = 7\cos t - 2\sin t$,
then $p\sin q = 7$ and $p\cos q = -2$ *[1 mark for both]*
$(p\sin q)^2 + (p\cos q)^2 = 7^2 + (-2)^2$
$\Rightarrow p^2\sin^2 q + p^2\cos^2 q = 49 + 4$
$\Rightarrow p^2(\sin^2 q + \cos^2 q) = 53 \Rightarrow p^2 = 53 \Rightarrow p = \sqrt{53}$ *[1 mark]*
$(p\sin q) \div (p\cos q) = 7 \div (-2) \Rightarrow \tan q = -\frac{7}{2}$
$\tan^{-1}\left(-\frac{7}{2}\right) = -1.292...$
$p\sin q = 7 \Rightarrow \sin q$ is positive, $p\cos q = -2 \Rightarrow \cos q$ is negative
Using an ASTC diagram, this means that $\frac{\pi}{2} < q < \pi$
So $q = -1.292... + \pi = 1.849...$ *[1 mark]*
So $A(t) = \sqrt{53}\sin(t + 1.849...)$
[4 marks available in total — as above]

b) When $A(t) = 1$, $\sqrt{53}\sin(t + 1.849...) = 1$ *[1 mark]*
$\Rightarrow \sin(t + 1.849...) = \frac{1}{\sqrt{53}}$
If $0 \le t \le 2\pi$, then $1.849... \le (t + 1.849...) \le 8.132...$ *[1 mark]*
$t + 1.849... = \sin^{-1}\left(\frac{1}{\sqrt{53}}\right) = 0.1377...$ (outside valid range)
Other solutions are at:
$t + 1.849... = \pi - 0.1377... = 3.003...$ (within valid range)
$t + 1.849... = 2\pi + 0.1377... = 6.420...$ (within valid range)
[1 mark for both]
Then $t = 3.003... - 1.849... = 1.15$ (3 s.f.)
and $t = 6.420... - 1.849... = 4.57$ (3 s.f.) *[1 mark for both]*
[4 marks available in total — as above]

12 a) Since **r**, **s** and **t** form the sides of a cuboid,
they must all be perpendicular to each other.
$\mathbf{r}.\mathbf{s} = 0 \Rightarrow (3\mathbf{i} + 2\mathbf{j} + 2\mathbf{k}).(2\mathbf{i} + m\mathbf{j} + 4\mathbf{k}) = 0$ *[1 mark]*
$\Rightarrow (3 \times 2) + (2 \times m) + (2 \times 4) = 0$
$\Rightarrow 6 + 2m + 8 = 0 \Rightarrow m = -14 \div 2 = -7$ *[1 mark]*
$\mathbf{r}.\mathbf{t} = 0 \Rightarrow (3\mathbf{i} + 2\mathbf{j} + 2\mathbf{k}).(n\mathbf{i} - 8\mathbf{j} - 25\mathbf{k}) = 0$ *[1 mark]*
$\Rightarrow (3 \times n) + (2 \times (-8)) + (2 \times (-25)) = 0$
$\Rightarrow 3n - 16 - 50 = 0 \Rightarrow n = 66 \div 3 = 22$ *[1 mark]*
[4 marks available in total — as above]
You could also have used the fact that s.t = 0.

b) $\overrightarrow{CP} = \frac{1}{2}\mathbf{t} = \frac{1}{2}(22\mathbf{i} - 8\mathbf{j} - 25\mathbf{k}) = 11\mathbf{i} - 4\mathbf{j} - 12.5\mathbf{k}$ *[1 mark]*
$\overrightarrow{AP} = \overrightarrow{AB} + \overrightarrow{BC} + \overrightarrow{CP} = \mathbf{r} + \mathbf{s} + \frac{1}{2}\mathbf{t}$
$= (3\mathbf{i} + 2\mathbf{j} + 2\mathbf{k}) + (2\mathbf{i} - 7\mathbf{j} + 4\mathbf{k}) + (11\mathbf{i} - 4\mathbf{j} - 12.5\mathbf{k})$
$= 16\mathbf{i} - 9\mathbf{j} - 6.5\mathbf{k}$ *[1 mark]*
[2 marks available in total — as above]

c) If angle BAP $= \theta$, then $\overrightarrow{AB}.\overrightarrow{AP} = |\overrightarrow{AB}| \times |\overrightarrow{AP}| \times \cos\theta$
$|\overrightarrow{AB}| = \sqrt{3^2 + 2^2 + 2^2} = \sqrt{9 + 4 + 4} = \sqrt{17}$ *[1 mark]*
$|\overrightarrow{AP}| = \sqrt{16^2 + (-9)^2 + (-6.5)^2}$
$= \sqrt{256 + 81 + 42.25} = \sqrt{379.25}$ *[1 mark]*
$\overrightarrow{AB}.\overrightarrow{AP} = (3\mathbf{i} + 2\mathbf{j} + 2\mathbf{k}).(16\mathbf{i} - 9\mathbf{j} - 6.5\mathbf{k})$
$= (3 \times 16) + (2 \times (-9)) + (2 \times (-6.5))$
$= 48 - 18 - 13 = 17$ *[1 mark]*
So $17 = \sqrt{17}\sqrt{379.25}\cos\theta$ *[1 mark]*
$\Rightarrow \cos\theta = 0.2117... \Rightarrow \theta = 77.8°$ (3 s.f.) *[1 mark]*
[5 marks available in total — as above]

13 The line and the curve intersect when $3x^2 - 2x^3 = -2x$ *[1 mark]*
$\Rightarrow 2x^3 - 3x^2 - 2x = 0 \Rightarrow x(2x^2 - 3x - 2) = 0$
$\Rightarrow x(2x + 1)(x - 2) = 0 \Rightarrow x = -\frac{1}{2}, 0$ or 2 *[1 mark]*
Integrate $(3x^2 - 2x^3) - (-2x)$ between $x = -\frac{1}{2}$ and $x = 0$,
and between $x = 0$ and $x = 2$, separately:
$\int_{-\frac{1}{2}}^{0} (3x^2 - 2x^3 + 2x)\,dx = \left[x^3 - \frac{1}{2}x^4 + x^2\right]_{-\frac{1}{2}}^{0}$
*[2 marks for the correct integral, otherwise
1 mark for at least two terms correct]*
$= \left((0)^3 - \frac{1}{2}(0)^4 + (0)^2\right) - \left(\left(-\frac{1}{2}\right)^3 - \frac{1}{2}\left(-\frac{1}{2}\right)^4 + \left(-\frac{1}{2}\right)^2\right)$
$= (0 - 0 + 0) - \left(-\frac{1}{8} - \frac{1}{32} + \frac{1}{4}\right) = 0 - \left(\frac{3}{32}\right) = -\frac{3}{32}$ *[1 mark]*
So the shaded area between $x = -\frac{1}{2}$ and $x = 0$ is $\frac{3}{32}$ units^2.
$\int_{0}^{2} (3x^2 - 2x^3 + 2x)\,dx = \left[x^3 - \frac{1}{2}x^4 + x^2\right]_{0}^{2}$
$= \left((2)^3 - \frac{1}{2}(2)^4 + (2)^2\right) - \left((0)^3 - \frac{1}{2}(0)^4 + (0)^2\right)$
$= (8 - 8 + 4) - (0 - 0 + 0) = 4 - 0 = 4$ *[1 mark]*
So the shaded area between $x = 0$ and $x = 2$ is 4 units^2.
Therefore, the total shaded area is $4 + \frac{3}{32} = 4\frac{3}{32}$ units^2. *[1 mark]*
[7 marks available in total — as above]

Answers

Formula Sheet

Revising for your Higher exams is tricky enough, without having to worry about learning more formulas than you absolutely need to. Luckily, this page shows you all the ones that you'll be given at the front of your exam papers. Make sure you know what each of them means and how to use them.

Circles

$x^2 + y^2 + 2gx + 2fy + c = 0$ represents a circle with centre $(-g, -f)$ and radius $\sqrt{g^2 + f^2 - c}$

$(x - a)^2 + (y - b)^2 = r^2$ represents a circle with centre (a, b) and radius r

The Scalar Product

$\mathbf{a}.\mathbf{b} = |\mathbf{a}||\mathbf{b}| \cos q$, where q is the angle between vectors $\mathbf{a}$ and $\mathbf{b}$

or $\mathbf{a}.\mathbf{b} = a_1 b_1 + a_2 b_2 + a_3 b_3$, where $\mathbf{a} = \begin{pmatrix} a_1 \\ a_2 \\ a_3 \end{pmatrix}$ and $\mathbf{b} = \begin{pmatrix} b_1 \\ b_2 \\ b_3 \end{pmatrix}$

Trig Formulas

$\sin(A \pm B) = \sin A \cos B \pm \cos A \sin B$

$\cos(A \pm B) = \cos A \cos B \mp \sin A \sin B$

$\sin 2A = 2 \sin A \cos A$

$\cos 2A = \cos^2 A - \sin^2 A$
$\qquad = 2 \cos^2 A - 1$
$\qquad = 1 - 2 \sin^2 A$

Standard Trig Derivatives

$f(x)$	$f'(x)$
$\sin ax$	$a \cos ax$
$\cos ax$	$-a \sin ax$

Standard Trig Integrals

$f(x)$	$\int f(x)\,dx$
$\sin ax$	$-\dfrac{1}{a} \cos ax + C$
$\cos ax$	$-\dfrac{1}{a} \sin ax + C$